Selected Titles in This Series

D1288643

Mathematical and Computational Biology:

Computational Morphogenesis, Hierarchical Complexity, and Digital Evolution

Lectures on Mathematics in the

LIFE SCIENCES

Volume 26

Mathematical and Computational Biology:

Computational Morphogenesis, Hierarchical
Complexity, and Digital Evolution

An International Workshop
21–25 October 1997
University of Aizu, Aizu-Wakamatsu City, Japan

Chrystopher L. Nehaniv
Editor

American Mathematical Society
Providence, Rhode Island

Proceedings of the International Workshop on Mathematical and Computational Biology, held at the University of Aizu, Aizu-Wakamatsu City, Japan, October 21–25, 1997.

Cover art created and provided by Richard Michod and Denis Roze.
Used with permission.

1991 *Mathematics Subject Classification*. Primary 92–06, 92–02, 92B05, 92D15.

Library of Congress Cataloging-in-Publication Data
International Workshop on Mathematical and Computational Biology (1997 : University of Aizu)
Mathematical and computational biology : computational morphogenesis, hierarchical complexity, and digital evolution: an International Workshop on Mathematical and Computational Biology, October 21–25, 1997, University of Aizu, Aizu-Wakamatsu City, Japan / Chrystopher L. Nehaniv, editor.
p. cm. — (Lectures on mathematics in the life sciences, ISSN 0075-8523 ; v. 26)
Includes bibliographical references.
ISBN 0-8218-0941-5 (alk. paper)
1. Biology—Mathematical models—Congresses. 2. Biology—Computer simulation–Congresses. I. Nehaniv, Chrystopher L., 1963– . II. Title. III. Series.
QH323.5.I5673   1997
570'.1'51—dc21
                                                                                98-54449
                                                                                CIP

Contents

Foreword

This volume contains thirteen lectures of the INTERNATIONAL WORKSHOP ON MATHEMATICAL & COMPUTATIONAL BIOLOGY, subtitled "Computational Morphogenesis, Hierarchical Complexity & Digital Evolution", held 21-25 October 1997 at the University of Aizu in Japan during some beautiful autumn days. This interdisciplinary international workshop brought together researchers working on aspects of evolutionary, mathematical and computational biology that are of particular interest (1) for computer scientists, both as sources of ideas (via emulation of nature's solutions) for software systems and as areas of application (bioinformatics), (2) for biologists using or looking for mathematical and computational methods to address difficult questions of evolution such as individuality, multicellularity, replication, and morphogenesis, and (3) for mathematicians who strive to develop the theoretical tools for asking (and possibly help answering!) such questions rigorously.

The invited plenary speakers, who provided excellent talks and/or papers, included Eörs Szathmáry, Larry Bull, Richard E. Michod, John L. Rhodes, Karl Sigmund, Thomas S. Ray, Kurt Fleischer, Joel R. Peck, and Giuseppe Pirillo. Jason D. Lohn of NASA gave an invited tutorial on self-replicating systems. Discussion panels at the workshop addressed the topics of "Origins, Maintenance, Proliferation and Diversity of Differentiating Self-Replicators" (L. Bull and T. S. Ray) and "Is There Something Wrong with the Mathematics of Morphogenesis & Evolutionary Biology?" (K. Fleischer, C. Nehaniv, K. Sigmund).

The lectures in this volume are roughly ordered according to a general increase of biological scale. Of special concern are self-replication, the evolution of individuality, symbiogenesis, evolutionary developmental biology, computational morphogenesis, the evolution and maintenance of sex, interaction dynamics within evolving populations, and properties of the digital genetic code.

Self-replicating and self-maintaining systems are providing models for advances in computational sciences, while the mathematical foundations of computer sciences are exerting a new influence in biology. E Szathmáry offers us a revised classification of replicators that includes phenotypic replicators such as prions and memes, while J D. Lohn covers recent developments in self-replicators in cellular spaces, an area of study initiated by John von Neumann.

L. Bull reveals conditions on fitness landscapes which may favor maintenance of symbiogenesis and of differentiated multicellularity. Deeper insights into the genesis of metazoan life will rely on mathematical and computational models which can elucidate synthesis of hierarchical structure while coping with the tendency of lower levels to evolve out-of-control (as happens in cancer but also in large software systems). This is addressed directly in the pioneering paper of R. E. Michod and D. Roze. Major transitions in evolution of living systems are now understood to have been brought about by changes in hierarchical structuring, resulting in new units of evolution, and in the way information is organized, manipulated, and interpreted. C. L. Nehaniv and J. L. Rhodes present a rigorous mathematical way of treating the complexity of biological systems and some consequences for rates of evolution.

Computational morphogenesis is treated in the papers by T. Unemi and by V. V. Savchenko, A. G. Basnakian and A. A. Pasko from the perspectives of evolution of development and of simulating colonial growth, respectively. The lecture by J. R. Peck, J. M. Yearsley and D. Waxman illuminates some previously mysterious aspects of the distribution of sexual vs. self-fertilizing and asexual populations. S. Okuyama offers a novel medical viewpoint on the evolutionary aspects of disease. The lecture by K. Sigmund studies the evolutionary dynamics of social agents modelled as automata engaged in the games of life. L. M. Schmitt and C. L. Nehaniv analyze the geometry and long-term behaviour of genetic algorithms using methods of linear algebra and real analysis. Intriguing combinatorial properties of the genetic code are discussed by G. Pirillo, and Q.-P. Gu, S. Peng and Q.-M. Chen apply the sorting of permutations to genomic analysis.

Although approaches to the nature, study and emulation of life through mathematical representation and digital media in synthetic worlds have progressed rapidly in the last decades, the development of mathematical frameworks for addressing these issues still remains largely a future prospect that we sought to address. The complexity of hierarchical systems can be treated for example using methods from the algebraic theory of automata. Models of how such higher levels of complexity arise in nature will have implications for the future of complex information systems as well as for the understanding of biological systems.

Some notion of a constructive dynamical system in which objects interact to produce other objects may be a key ingredient in an adequate general mathematical treatment of living systems. These will complement more traditional uses of differential models circumscribe a fixed set of abstracted, measured quantities. Means of introducing new qualities and quantities are elusive aspects of biological evolution and development that our mathematics must eventually address. We are witnessing a steady rise in combinatorial, algebraic, automata-theoretic and other computational approaches in mathematical and computational biology as they attempt to meet these challenges of understanding, modelling, simulating, building, and predicting as we come ever closer to the essential features of life itself. There is a new spirit of *Constructive Biology* emerging from the synthesis of mathematical, computational and biological perspectives. This area seeks understanding by concentrating on what is sufficient or necessary to *construct* a system exhibiting structure, properties, or behaviour of a biological system, and on applying this knowledge in new settings.

Thus, the workshop sought to provide a multi-disciplinary forum in which researchers from various fields could discuss and develop the ideas relating biology, symbiogenesis, autopoiesis, self-reproducing and self-maintaining systems, constructive biology, computational morphogenesis, biological cybernetics, animation using digitally evolved actors, robotics, embodied agents, mathematical models in ecology and population genetics, mathematical foundations of theoretical computer science including concurrency, parallelism, λ-calculus, dynamical systems and semigroup theory, dynamic constructors, formal languages, automata, as well as future applications to hierarchical software systems and synthetic worlds. Also methods, models and applications of digital evolution were a central theme.

The workshop program was interrupted for a visit to the traditional Japanese mountain hot-springs, enjoyment of the local cuisine and *sâke*, and a tour of the old castle town of Aizu-Wakamatsu City including a visit to the 17th century double-helical(!) Buddhist temple on historic Mt. Iimoriyama.

We gratefully acknowledge all the speakers and conference participants for the important roles they each played in making the workshop such a success.

Along with the accepted contributed papers, all invited papers appearing herein have been fully peer-refereed. We thank the anonymous referees whose generous review work was invaluable to the production of this volume.

We gratefully acknowledge the generous support, facilities and sponsorship of the workshop provided by the University of Aizu. In particular, we thank for their time and assistance the International Affairs Committee, Prof. Charles Moore, Prof. Nikolay Mirenkov, Mr. Meike, the Academic Liason Office, Mr. Kuniyasu Izumi, Mrs. Mannching Wang, Mrs. Galina Pasko, Mrs. Oba, Prof. Minetada Osano, Prof. Masami Ito, colleagues at the University of Aizu, students of the Software Engineering Lab and of the Cybernetics and Software Systems Group. Prof. Richard E. Michod kindly provided the cover graphic suggesting conflict and cooperation in a possible scenario for transition to differentiated mullicellular individuality in eukaryotes.

We are pleased to acknowledge the kind cooperation of the American Mathematical Society, and additional support from University of Aizu projects "Algebra & Computation" (R-10-1) and "Bioinformatics" (G-24). The University of Hertfordshire, in particular, Prof. Martin Loomes and the Interactive Systems Engineering Research Group, provided time and facilities for completing work on this volume.

Finally, we are pleased to thank the kind people of Aizu-Wakamatsu City for their wonderful hospitality.

Chrystopher L. Nehaniv
October 1998

University of Aizu
Japan

University of Hertfordshire
United Kingdom

Lectures on Mathematics in the Life Sciences
Volume **26**, 1999

Chemes, Genes, Memes: A Revised Classification of Replicators

Eörs Szathmáry

ABSTRACT. A revised classification of replicators is presented. The two largest categories are genotypic and phenotypic replicators. In the case of the latter only certain phenotypic traits are transmitted, whereas in the former the genotype is passed on. The replication process can be holistic or modular. Hereditary potential can be limited or unlimited. Examples from chemistry and biology abound. The only empty class is holistic replicators with unlimited heredity. Presumably, the typical path of evolution proceeds from limited to unlimited heredity and from holistic to modular replication.

1. Introduction

Ever since Dawkins [10] introduced the concept of a replicator, people have become used to thinking of genes and memes as replicators. Various definitions of replicators hinge on the notions of transmission of structure and of accurate copying [24]. In this paper, I will show that there are several other kinds of replicator, and not all of them rest on modular copying of a certain structure. In the case of genetic replicators, heredity of the phenotype rests on copying (replication) of the genotype. In other cases the phenotype is transmitted through a direct process, different from classical template replication. Curiously enough, this applies also to memes. Direct transmission of the phenotype allows for 'Lamarckian' (or, more properly, non-Weismannian) evolution. Even in those cases where replication rests on the direct copying of the genotype, it is not necessarily digital. In fact, the earliest replicators are likely to have been holistic autocatalysts, undergoing processive, rather than modular, replication. I present the examples according to the logic of a revision of an earlier classification [42]. Finally, I present a particularly clear example of a reproducer in the form of the chemoton model [19-21], so far the best abstraction of minimal life.

1991 *Mathematics Subject Classification*. Primary 92B05, 92D15; Secondary 92C40.
The author was supported by the Hungarian Scientific Research Fund (OTKA).

2. Genotypic replicators

As Leslie Orgel [36] noted, in any chemical system replication rests on autocatalysis and autocatalysis always results, in some sense, in replication. (An autocatalyst is a compound that catalyses its own formation from raw materials.) This definition is important, since the notion of catalysis already implies cycles of operation, which can be repeated indefinitely. Note also that it does not restrict the mechanism to a modular copying process or digital information storage. Following Orgel [36, 37], we are interested in informational replication, or hereditary replicators (see also [33, 43]). It may be thought that nucleic acids are the only hereditary replicators, but this is manifestly not so, as we shall see in a moment. In short, there are different "senses" of replication. I suggest that, *following the analogies of the gene and the meme, we can refer to any molecular replicator as a cheme.*

2.1. Replicators with limited heredity. Limited heredity is context-dependent, since we require that the number of possible types should be smaller than the number of actual individuals present [42]. Although this definition is context-dependent, FAPP (for all practical purposes) it is readily applicable. For example, an imaginary system of replicating hexadeoxynucleotides (or their analogues) will have limited heredity in a test tube containing 10^5 molecules, since there are only $4^6 = 4096$ possible sequences. Note that a chemical system of 10^5 molecules is a very small one, if one thinks of Avogadro's number ($6*10^{23}$). Thus in almost all realistic systems hexanucleotides will qualify as limited hereditary replicators.

2.1.1. *Holistic replicators with limited heredity.* As early as 1861, Butlerow discovered the first reaction of potential prebiotic importance. It is called the formose reaction, and is in fact a complex network of interconversions of different sugar

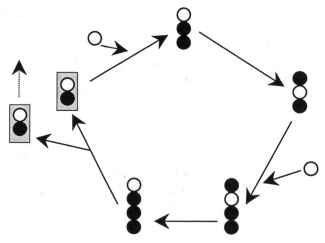

FIGURE 1. The autocatalytic core of the formose reaction. Circles represent groups with one carbon atom [44]. Solitary circle: fomaldehyde. Framed object: glycolaldehyde.

molecules of different size (see e.g. [4]). One can start the system with an alcalic solution of formaldehyde. After a lag period the first sugar molecules (trioses) appear, and the concentration of sugars increases exponentially, indicating that some autocatalysis must be going on. The core of the reaction is the autocatalytic formation of glycolaldehyde from formaldehyde (Figure 1).

As far as we know, replication in the formose system is entirely non-informational, there are no hereditary variants. Nevertheless, it is instructive to look at the mechanism of replication. It *does not make sense to say that replication of a glycolaldehyde is halfway through, since the carbon groups are not copied one by one.* That is why we call them analogue [42] or holistic [34] replicators. The second important feature of the system is that it operates without enzymatic aid. This is not true of other examples, such as the Calvin cycle or the reductive citric acid cycle: these are different autocatalytic cycles fixing carbon dioxide in contemporary organisms. In contrast to the formose system, specific enzymes catalyze each elementary reaction.

Can thus there be holistic replicators with real heredity at all? We do not know, but it may be possible. For example, Wächtershäuser [47, 48] suggested alternative variants of an archaic reductive citric acid cycle. King [29] hypothesized that coenzymes were part of pre-enzymatic, autocatalytic chemical networks. Before and after inclusion of a particular coenzyme we have different, but related networks. It is remarkable that Eakin argued in 1963 that coenzymes could have been the earliest replicators [15].

As Wächtershäuser [46, 48, 49] argued, the emergence of a chemically novel compound and its grafting onto the network will result in a so-called *memory effect,* the generalized chemical basis of heredity. He presents three examples of extensions to autocatalytic cycles (Figure 2): real hereditary changes will be due to *vitalysts* (Figure 2b) and *virulysts* (Figure 2c). The former is a compound that catalyses the operation of

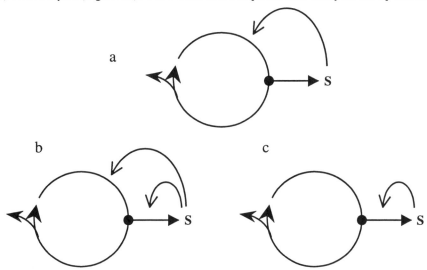

FIGURE 2. The product S of a side-reaction of an autocatalytic cycle can feed back catalytically on various parts of the network. (a) altruyst; (b) vitalyst; (c) virulyst. Modified from [49].

the cycle as well as the branch leading to its own formation. The latter compound selfishly catalyses only the (parasitic) branch leading to itself. It is the vitalysts that can be successfully grafted onto the pre-existing network: certain coenzymes in the primordial world could have been added to metabolism as real vitalysts [29].

King advocated another possibility of chemical complexification, namely chemical symbiosis of different cycles [29]. This could happen with addition, subtraction, or simple juxtaposition of the two networks (see also [48, 17]). Subtraction means that some parts of the networks to be united will be discarded, addition means that some chemical "glue" will be added between the two systems. Subtraction and addition must, of course, arise spontaneously, as a result of the chemical affinities of the participating molecules.

I must emphasize that, although these examples are fascinating, they are hypothetical. We do not know if holistic replicators can have heredity. If they can, their hereditary changes are analogous to "macromutations", and the evolutionary process they undergo is macroevolution, rather   than microevolution [42]. This seems to be connected to their replication being holistic. The successful establishment of a vitalystmust be a very rare event [49]. But it may also turn out, experimentally, that holistic replicators do not have heredity. This would pose severe difficulties for the origin of life (cf. [33, 34]). What seems quite certain, due to the processive nature of replication, is that holistic replicators will not show unlimited heredity. Due to the lack of modular replication, most aspects of phenotypic traits (concentration fluctuations in the cycle, etc.) will not be transmitted.

As a referee has pointed out, it is certainly possible to conceive of replicators in which concentrations of various chemicals kept, say, in solutions in separate compartments, could be inherited. The compartments might be possibly linearly arranged, so that sequences of compartments might have arbitrary length. Admittedly, no such self-replicating chemical system is known. Nevertheless, if it did exist, would it be an example of a non-digital genotypic system? I do not think so: the linear arrangement of compartments would lend itself to a modular replication process; it would certainly make sense to say that replication of the array is half complete.

2.1.2. *Digital replicators with limited heredity.* Many examples belong to this category, including the first non-enzymatically replicating hexadeoxynucleotide analogue synthesized by von Kiedrowski [28]. Rebek *et al.* constructed a system, which is only very remotely similar to nucleic acids [23]. Note that the end of the replicating J molecule (Figure 3) can be varied, thus different (geno)types of replicator can be made. A modification of von Kiedrowski's original scheme also yielded limited heredity [2].

The foregoing may seem to suggest that replicators are necessarily one-dimensional. This is not so; we shall see the example of membranes later. Clay minerals, if they could really be replicating [5], would belong to this category. A hypothetical, artificial example was suggested by Orgel [37].

The interesting fact is that all digital replicators with limited heredity are artificial compounds so far. Thus they have a strong didactic value, but have no direct relevance to the problem of early evolution. Several people have hypothesized [14, 46, 48] that modular replicators appeared as by-products of primordial metabolism. If so, then evolution proceeds from holistic to modular and from limited to unlimited hereditary replicators [42, 44]. This brings us to the next non-empty category.

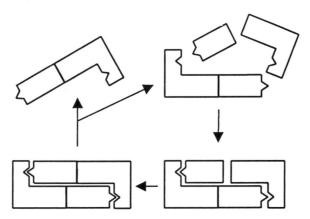

FIGURE 3. Scheme of an artificial replicator. Note that by varying the complementary shapes at the two ends of the J molecule a series of different replicators can be made [23].

2.1.3. *Digital replicators with unlimited heredity.* The commonly known replicators (viruses, genes, asexual genomes, etc.) belong to this category. Although the set of possible sequences is not infinitely large in the mathematical sense, FAPP the search through sequence space [32] can go on indefinitely. Imagine a vast sequence space accommodating all possible nucleotide sequences up to 10^9 nucleotides (the size of the human genome), which is initially completely dark. Suppose you would sequence all extant individual genomes, and you would light a bulb at each corresponding point of the sequence space. Having completed this formidable task, the space would still remain essentially dark. The power of unlimited heredity is exemplified by not only the existing genes, but also by their artificial counterparts. The power of *in vitro* genetics has produced molecules with predetermined function by replication and unnatural selection [37].

The chromatin-marking system (by either methylation or proteins) in multicellular eukaryotic development is the most important epigenetic inheritance system [25], allowing for a large number of cell types in the organism with the same genome. It is powerful because it is digital [26].

3. Phenotypic replicators

This is an entirely new category, never discussed before. Phenotypic replication is a process whereby the phenotype, or function, of one object is transmitted to the other, without any modular copying effect. The following process can illustrate it. Suppose you were choosing your offspring from a large pool of babies according to some criterion based on physical and/or emotional resemblance. Or, in a more extreme imaginary case, you would watch babies emerging from a soup and you would catalyze somehow the formation of babies resembling you in some respect. Although phenotypic replication is a new concept, it is striking to realize that it applies to memes as introduced by Dawkins [10].

3.1. Phenotypic replicators with limited heredity. Prions and genetic membranes qualify as examples; I shall discuss them in turn. Prions seem to replicate without a nucleic acid component. One suggested mechanism is the transfer of a certain conformation (i.e. phenotypic trait) from one molecule to another (Figure 4). The prion molecule from humans has two sequences *A* and *B*, and two conformations, *a* and *b*. Both sequences can adopt both conformations [35]. If, by definition, *b* is the harmful phenotype, then *B* sequences have a higher chance to adopt this conformation. Suppose that a *B* sequence with conformation *b* gets in contact with an *A* protein in conformation *a*. What happens is that the conformation of *A* will change to *b*. The sum of process is the following:

$$A + B \rightarrow A + B,$$
$$a + b \rightarrow 2\,b.$$

This clearly shows that a phenotypic trait is passed on. Hence this field has been referred to as "molecular phenetics" [38]. At the heart of the process lies a "memory effect" of altered protein folding [40].

The transmission of phenotype is the essential 'Lamarckian' component of cultural evolution [34], rather than the hopping of a meme from one individual to the next [16], since the latter case is no different from that of a virus going from host to host [38].

It seems that prion inheritance has been hijacked by organisms a number of times. Certain phenotypic states in yeast are "coded for" by prions [39, 45], and a new prion controls fungal cell fusion incompatibility [50]. Thus it seems that prions have occasionally been recruited by organisms as epigenetic inheritance systems [26].

The category of genetic membranes is due to Cavalier-Smith [7]. It refers to membranes, which within their contemporary environments cannot be made *de novo*. Bacterial membranes (plasma and pigment membranes), membranes of the endoplasmic reticulum and the nucleus, those of mitochondria, plastids and microbodies belong to this category. These membranes always grow and divide, but are never created *de novo*. The reason is that some of the information to create them resides in the membrane itself. Take the example of plastids and mitochondria. Most of their proteins are coded for by the nucleus and must be imported from the cytoplasm. Proteins destined to get into these organelles carry an N-terminal signal sequence, which folds into an appropriate conformation. This conformation, rather than the sequence itself, is recognized by the

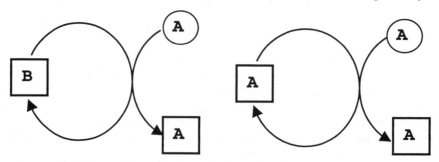

FIGURE 4. Phenotypic replication of prions. The box and the circle refer to the harmful (*b*) and to the normal (*a*) conformation, respectively.

transfer system (receptor) in the corresponding membranes (in fact the sequences themselves are not conserved). One could completely compromise the system by exchanging the receptors in the two kinds of membrane, without any change in either the nuclear, mitochondrial, or plastid genomes.

The autocatalytic part becomes clear if one realizes that the specific receptors are made of proteins, which also must be taken up from the cytoplasm if they are not to be diluted out through organelle growth and division. Although the autocatalytic nature of membrane growth was realized a long time ago [18], the hereditary nature of some membranes deserves special attention.

Another kind of epigenetic inheritance system is the cortical (structural) inheritance in ciliates, originally described by Sonneborn [41]. It also is a two-dimensional phenotypic inheritance system, whereby the orientation of kinetids (basal body plus associated cortical and fibrillar structures) is passed on from cell generation to generation. Experimental demonstration of micelle and vesicle replication [1], as well as self-replicating amphiphilic monolayers [31] adds to the examples of phenotypic replicators in simple chemical systems.

Memes of nonhuman species (e.g. certain bird songs) also belong to this category.

3.2. Phenotypic replicators with unlimited heredity. The only example I can think of at the moment is human memes (memes cultivated by human brains) [10, 11, 3]. Considering the apparently unlimited variation in cultural evolution, one cannot deny unlimited heredity of the memes in humans. This is associated with the human language, whereby an indefinitely large number of grammatical sentences can be constructed [27, 33].

Cloak distinguished between i-culture and m-culture [8], which Dawkins equated with the meme as replicator in the brain and the effect of the meme on the external world, respectively [11]. Dawkins [11] suggested that the meme should be thought of as a unit of the information content *in the brain*.

It is instructive to look into the details of memetic replication. When one teaches Newton's second law to students, it is the end result that is similar to one's initial state: the teacher and the student will both be able to solve problems when knowledge of that law is needed. Verbal recapitulation is not taken as sufficient: the mere verbal formulation (like that of a limerick) should be taken as a different meme. From what we know about the function of brain areas (e.g. [12]) it is probably true that the underlying neural structures "coding for" the same meme in different brains will usually not be recognisably similar. It is only in the context of all the other neural structures, resulting from personal neural and mental history, that they have the same effect, i.e. phenotype. This is, I repeat, why memes are phenotypic replicators, and cultural evolution is 'Lamarckian' [34].

Another question is the possibility of memetic replication inside the brain. If it turned out to be real, it would serve as an ideal substrate for "neuronal Darwinism", *sensu stricto* [9]. Calvin demonstrated that it is at least theoretically possible to design replicative patterns of neural activity, which could undergo evolution by natural selection within the brain [6]. The "cerebral code" thus established would be digital [6]; as one would expect from the experience that holistic replicators have only limited heredity (see above), in contrast to the hereditary potential of memes. Note, however, that this within-brain replication would be genotypic, rather than phenotypic.

But this is all speculative at the moment. We do not even know whether discrete abstract entities, such natural numbers, are encoded digitally or not in animal brains [13].

4. Design for a Reproducer

As Griesemer [22] argued, cells and organisms are not replicators *sensu* Dawkins [10], but they show heredity and they reproduce. Reproduction involves replication of the genes as well as development of the organism. (The concept of a reproducer is reviewed in Ref. **44**.)

It may be, at first sight, striking that cells are not replicators, since typically all their major component systems (cytoplasm, genetic material and cell envelope) are. In fact, the most elegant model for minimal life, proposed by Gánti, emphasizes exactly this observation. His so-called chemoton model is an autocatalytic chemical super-system composed of a metabolic, a genetic, and a boundary subsystem (Figure 5), each of which is autocatalytic, and therefore replicates [18-21]. Is, then, the chemoton as a whole a replicator or not?

This paradox can, I think, be resolved by the application of the concepts outlined in this paper. As we have seen, phenotypic replicators pass on only certain

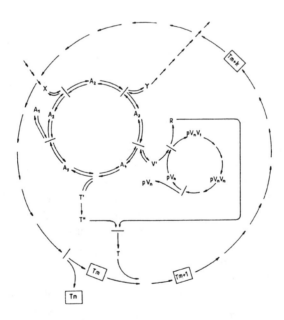

FIGURE 5. The design of Gánti's chemoton. Molecules A_I are the intermediates of the metabolic cycle; V' is the activated template monomer; pV_n is the template polymer consisting of n monomers; R is a condensation by-product; T* is the membranogenic molecule; T_m is a membrane built of m pieces of T [21]. Reprinted from "Biogenesis itself" by Tibor Gánti in *Journal of Theoretical Biology* (187), pp. 583–593, 1997, by permission of the publisher Academic Press, London, UK.

aspects of their phenotype, and this applies to the membrane subsystem. Intermediates of the metabolic cycle are holistic replicators. Both subsystems have at most limited heredity, and most aspects of their phenotypes are not passed on accurately. Depending on the concrete biochemistry, these two subsystems, although autocatalytic, may just be non-informational replicators *sensu* Orgel. This is why the chemoton as a whole is not a replicator *sensu* Dawkins. I suggest, complementing Griesemer's penetrating analyses, that *reproducers are composite systems of different kinds of replicators, some of which have only (at most) limited heredity.*

5. References

1. Bachmann, P. A., Luisi, P. L. & Lang, J. *Autocatalytic self-replicating micelles as models for prebiotic structures,* Nature **357** (1992), 57-59.
2. Achilles, T. & von Kiedrowksi, G., *A self-replicating system from three starting materials.* Angew. Chem. Int. Ed. Engl. **32** (1993), 1198-1201.
3. Ball, J. A., *Memes as replicators,* Ethology Sociobiol. **5** (1984), 145-161.
4. Cairns-Smith, A. G. & Walker, G. L. *Primitive metabolism,* BioSystems **5** (1974), 173-186.
5. Cairns-Smith, A. G., *Genetic Takeover and the Mineral Origin of Life,* Cambridge Univ. Press, Cambridge, 1982.
6. Calvin, W., *The Cerebral Code,* MIT Press, Cambridge, 1996.
7. Cavalier-Smith, T., In *Biodiversity and Evolution* (Kato, M. & Doi, Y. eds) pp. 75-114, National Science Museum Foundation, Tokyo, 1995.
8. Cloak, F. T., *Is a cultural ethology possible?* Human Ecol. **3** (1975), 161-182.
9. Cziko, G., *Without Miracles,* MIT Press, Mass., 1995.
10. Dawkins, R., *The Selfish Gene,* Oxford University Press, Oxford, 1976.
11. Dawkins, R., *The Extended Phenotype*, Freeman, Oxford, 1982.
12. Deacon, T. W., *The Symbolic Species,* W. W. Norton & Comp., New York, 1997.
13. Dehaene, S., *The Number Sense: How the Mind Creates Mathematics,* Oxford Univ. Press, Oxford, 1997.
14. Dyson, F. J. *Origins of Life,* Cambridge Univ. Press, Cambridge, 1985.
15. Eakin, R. E. *An approach to the evolution of metabolism,* Proc. Natl. Acad. Sci. USA **49** (1963), 360-366.
16. Edmunds, W. J. & Yool, A., *Is the propagation of prion molecules in different hosts an example of lamarckian inheritance?* Trends Ecol. Evol. **12** (1997), 194.
17. Fontana, W. & Buss, L. W., *"The arrival of the fittest": Toward a theory of biological organization.* Bull. Math. Biol. **56** (1994), 1-64.
18. Gánti, T. *Organization of chemical reations into dividing and metabolizing units: the chemotons.* BioSystems **7** (1975), 15-21.
19. Gánti, T. *A Theory of Biochemical Supersystems,* Akadémiai Kiadó, Budapest and University Park Press, Baltimore, 1979.
20. Gánti, T. *The Principle of Life,* OMIKK, Budapest, 1987.
21. Gánti, T. *Biogenesis itself,* J. theor. Biol. **187** (1997), 583-593.
22. Griesemer, J. *The Concept of a Reproducer* to be submitted.
23. Hong, J.-I., Feng, Q., Rotello, V. & Rebek, J. *Competition, cooperation, and mutation: Improving a synthetic replicator by light irradiation,* Science **255** (1992), 848-850.
24. Hull, D. L., *Individuality and selection.* Ann. Rev. Ecol. Syst. **11** (1980), 311-332.
25. Jablonka, E. & Lamb, M. J., *Epigenetic Inheritance and Evolution,* Oxford Univ. Press, Oxford, 1995.
26. Jablonka, E., Lamb, M. J. & Avital, E., *'Lamarckian' mechanisms in Darwinian evolution,* Trends Ecol. Evol. **13** (1998), 206-210.
27. Jablonka, E. & Szathmáry, E., *The evolution of information storage and heredity*, Trends Ecol. Evol. **10** (1995), 206-211.

28. Kiedrowski, G. von, *A self-replicating hexadeoxy nucleotide,* Angew. Chem. Int. Ed. Engl. **25** (1986), 932-935.
29. King, G. A. M. *Evolution of the coenzymes,* BioSystems **13** (1980), 23-45.
30. Lorsch, J. R. & Szostak, J. W. *Chance and necessity in the selection of nucleic acid catalysts,* Acc. Chem. Res. **29** (1996), 103-110.
31. Maoz, R., Matlis, S., DiMasi, E., Ocko, B. M. & Sagiv, J., *Self-replicating amphiphilic monolayers,* Nature **384** (1996), 150-153.
32. Maynard Smith, J. *Natural selection and the concept of a protein space,* Nature **225** (1970), 563-564.
33. Maynard Smith, J. & Szathmáry, E., *The Major Transitions in Evolution,* Freeman, Oxford, 1995.
34. Maynard Smith, J. & Szathmáry, E., *The Origins of Life.* Oxford University Press, Oxford, 1999.
35. Mestel, R. *Putting prions to the test,* Science **273** (1996), 184-189.
36. Orgel, L. E., *Molecular replication,* Nature **358** (1992), 203-209.
37. Orgel, L. E., *Unnatural selection in chemical systems,* Acc. Chem. Res. **28** (1995), 109-118.
38. Pagel, M. & Krakauer, D. C., *Prions and the new molecular phenetics,* Trends Ecol. Evol., **11** (1996), 487-488.
39. Patino, M. M., Liu, J.-J., Glover, J. R. & Lindquist, S. *Support for the prion hypothesis for the inheritance of a phenotypic trait in yeast.* Science **273** (1996) 622-626.
40. Shinde, U. P., Liu, J. J. & Inouye, M., *Protein memory through altered folding mediated by intramolecular chaperones,* Nature **389** (1997), 520-522.
41. Sonneborn, T. M., *The differentiation of cells,* Proc. Natl. Acad. Sci. USA, **51** (1964), 915-929.
42. Szathmáry, E., *A classification of replicators and lambda-calculus models of biological organization.* Proc. R. Soc. Lond. B **260** (1995), 279-286.
43. Szathmáry, E. & Maynard Smith, J., *The major evolutionary transitions.* Nature **374** (1995), 227-232.
44. Szathmáry, E. & Maynard Smith, J. *From replicators to reproducers: The first major transitions leading to life,* J. theor. Biol. **187** (1997), 555-571.
45. Tuite, M. F. & Lindquist, S. L., *Maintenance and inheritance of yeast prions,* Trends Genet. **12** (1996), 467-471.
46. Wächtershäuser, G. *Before enzymes and templates: theory of surface metabolism.* Microbobiol. Rev. **52** (1988), 452-484.
47. Wächtershäuser, G. *Evolution of the first metabolic cycles,* Proc. Natl. Acad. Sci. USA **87** (1990), 200-204.
48. Wächtershäuser, G. *Groundworks for an evolutionary biochemistry: the iron-sulfur world.* Prog. Biophys. Molec. Biol. **58** (1992), 85-201.
49. Wächtershäuser, G. *Vitalysts and virulysts: a theory of self-expanding reproduction.* In: Bentson, S. (ed) *Early Life on Earth. Nobel Symposium No. 84,* Columbia Univ. Press, New York, 1994, pp. 124-132.
50. Wickner, R. B. *A new prion controls fungal cell fusion incompatibility,* Proc. Natl. Acad. Sci. USA **94** (1997), 10012-10014.

DEPT OF PLANT TAXONOMY AND ECOLOGY, EÖTVÖS UNIVERSITY, LUDOVIKA TÉR 2, H-1083 BUDAPEST and COLLEGIUM BUDAPEST (INSTITUTE FOR ADVANCED STUDY), SZENTHÁROMSÁG U. 2, H-1014 BUDAPEST, HUNGARY
   E-mail address: szathmary@colbud.hu

Lectures on Mathematics in the Life Sciences
Volume **26**, 1999

Cellular Space Models of Self-Replicating Systems

Jason D. Lohn

ABSTRACT. Biological organisms are the most familiar examples of self-replicating systems, and until the late 1940s, the only instances formally researched. At that time, mathematicians and scientists began studying artificial self-replicating systems when it became desirable to gain a deeper understanding of how complex systems are able to form and evolve. Initial models consisted of abstract logical machines, or automata, embedded in cellular spaces. The large complexities seen in these early models agreed with the intuition that self-replication was an inherently complex process. Later, it was learned that much of the complexity was due to the imposition of artificial requirements. This paper traces developments from complex, early models of self-replicating systems in cellular spaces to recent, less complex models. As a survey of past models, this paper provides an overview of numerous self-replicating systems as well as some recent models that rely on emergent processes and artificial evolution.

1. Introduction

The brilliant mathematician John von Neumann initiated the formal study of artificial self-replicating systems in 1948, and before his untimely death in 1957, he had produced the first logical design of a self-replicating automaton [**31**]. Over the decades since this demonstration, theoretical and modeling studies have led to progressively simpler and smaller structures [**7, 3, 13, 23**]. They have produced structures that do problem solving while replicating [**6, 20, 27**], structures that were produced automatically via artificial evolution [**10, 16**], as well as demonstrated that self-replicating structures can emerge from a "sea" of non-replicating components [**5**]. The focus of this paper is to describe these and other models of artificial self-replication, limiting our attention to those embedded in cellular spaces. First we look at look at why these models are worthy of investigation and the historical trend away from complexity.

A better understanding of self-replicating systems could be useful in a number of ways, for both theoretical and practical purposes. Von Neumann was interested in understanding the build-up and evolution of extremely complex systems. Since biological organisms were known to be of enormous complexity and had complicated self-replication processes, he thought it natural to research self-replicating systems. He was especially interested in how a complex system could be constructed out of numerous simple parts. More recently, other reasons for studying

1991 *Mathematics Subject Classification.* Primary 68Q80, 92B05; Secondary 93A30, 58F08, 68Q15.

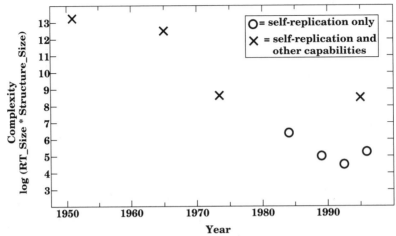

FIGURE 1. Plot of self-replicating system complexity for cellular automata models. From left to right, the × symbols are von Neumann [**31**], Codd [**7**], Vitányi [**29**], Perrier *et al.* [**20**], and the O symbols are Langton [**13**], Byl [**3**], Reggia *et al.* [**23**], Lohn and Reggia [**17**].

abstract models self-replication were posited. The field of artificial life was largely borne out of studies of such models. Subsequent researchers in artificial life found self-replicating structures to be a natural goal, especially for studying bottom-up, synthetic biologies. On a more practical bent, research on self-replicating structures could be useful in areas such as molecular-scale manufacturing [**9**], programming massively parallel computers [**22**], and computer virus research [**12**]. Researchers in molecular-scale manufacturing (also called nanotechnology), have discussed the potential of self-replicating systems: "If assemblers are to process large quantities of material atom-by-atom, many will be needed; this makes pursuit of self-replicating systems a natural goal." [**9**]. Having self-replicating computer programs that can be acted upon by digital evolution [**22**] could allow easier programming of massively parallel computers. Evolutionary bred self-replicating programs would breed on the parallel computer and the programs that most satisfy a set of requirements would be allowed to survive and replicate. Researchers have also investigated self-replicating structures to aid in understanding biomolecular mechanisms of reproduction and the origins of life [**11**].

Cellular space models of self-replicating systems have progressed from complex models to less-complex models. This trend is apparent in Figure 1, where complexity is plotted against time for cellular automata models. There are certainly other measures of complexity one could chose, but we have defined it to be the product of rule table size and structure size, plotted logarithmically. As can be seen, models designed only for self-replication are lowest in complexity, with the least complex of the others having three orders of magnitude more complexity.

The remainder of the paper is divided into three main sections. In section 2 we present background material regarding cellular space models and self-replicating structures. In section 3 we present a series of brief case studies beginning with the work of von Neumann and continuing through till present day. Included here

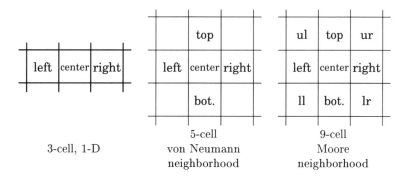

FIGURE 2. Common neighborhood patterns in 2-D cellular space models.

are models that are purely self-replicators, and those that provide additional functionality in addition to self-replication. In section 4 we summarize the paper and discuss potential research directions in the field.

2. Background

2.1. Cellular Space Models. A cellular space is a tessellation of cells containing finite state automata that interact with each other. The key properties of cellular space models are: strictly local interactions (resulting in emergent behavior), rule based automata (usually deterministic), high parallelism, simple automata (in general), and discretized space and time. Cellular automata (CA) models are the most widely studied models and constitute the majority of cellular space models. There are also numerous variations of the standard cellular automata model – for example: models that embed complex automata (e.g., a CPU with registers) within the cellular space, and models that allow other events or operators to act in addition to state transitions. These models are still cellular space models, yet it would not be proper to call them cellular automata.

2.2. Cellular Automata. Von Neumann co-invented cellular automata with Stanislaw Ulam as a medium in which to investigate and design complex systems such as self-replicating machines. Cellular automata are a class of spatially-distributed dynamical system models in which many simple components interact to produce potentially complex patterns of behavior [**7, 32**]. In a cellular automata model, time is discrete, and space is divided into an N-dimensional lattice of cells, each cell representing a finite state machine or automaton. All cells change state simultaneously with each using the same function δ or rule table to determine its next state as a function of its current state and the state of neighboring cells. This set of adjacent cells is called a *neighborhood*, the size of which, n, is commonly five or nine cells in 2-D models (see Figure 2). By convention, the center cell is included in its own neighborhood. Each cell can be in one of k possible states, one of which is designated the quiescent or inactive state. When a quiescent cell has an entirely quiescent neighborhood, a widely accepted convention is that it will remain quiescent at the next time step.

The CA rule table is a list of transition rules that specify the next state for every possible neighborhood combination. In a 2-D, 5-neighbor model the individual transition rules would be of the form CTRBL $\rightarrow$ C', where CTRBL specifies the

CTRBL	C'	CTRBL	C'	CTRBL	C'	CTRBL	C'
00000	0	01000	1	10000	1	11000	0
00001	1	01001	0	10001	0	11001	1
00010	1	01010	0	10010	0	11010	1
00011	0	01011	1	10011	1	11011	0
00100	1	01100	0	10100	0	11100	1
00101	0	01101	1	10101	1	11101	0
00110	0	01110	1	10110	1	11110	0
00111	1	01111	0	10111	0	11111	1

TABLE 1. Example CA rule table for the parity function.

states of the Center, Top, Right, Bottom, and Left positions of the neighborhood's present state, and C' represents the next state of the center cell.

The underlying space of CA models is typically defined as being isotropic, meaning that the absolute directions of north, south, east, and west are indistinguishable. However, the rotational symmetry of cell states is frequently varied. Strong rotational symmetry implies that all cell states are unoriented, meaning that each neighbor to a cell has no distinguishable position. Weak rotational symmetry implies that at least one cell state[1] is directionally oriented, meaning that the cell designates specific neighbors as being its top, right, bottom, and left neighbors. For example, the cell state designated ↑ in von Neumann's work is weakly-symmetric and thus permutes to different cell states →, ↓, and ← under successive 90° rotations. It represents one oriented *component* that can exist in four orientations. In CAs that contain both weak and strong rotationally symmetric states, it is common to represent the "strong" states using symbols that appear rotationally symmetric (e.g., ∘, +, ×), and the "weak" states (components) using symbols that are not rotationally symmetric (e.g., ↑, A, L).

As an example CA, consider the parity function where a cell's next state is one if there are an odd number of ones in the five cell neighborhood. The rule table for this function is shown in Table 1, and the configurations at four points in time are shown in Figure 3. It is interesting to note the complex patterns that arise from a simple two state, five neighbor function. Such dynamics illustrate the emergent behavior that is typical of many cellular automata simulations. Also note that CAs are typically very sensitive to initial conditions. For example, even a slight change to the $t=0$ state of the parity example will drastically change the dynamics.

2.3. Self-replicating structures. There is no universally accepted definition of a self-replicating structure, but we can qualitatively describe a generic self-replicating structure as follows. The structure itself is typically represented as a configuration of contiguous non-quiescent cells (see the five-component structure in Figure 4). As the space iterates, the structures goes through a sequence of steps to form a replicant. At some time t', copy of the original structure appears isolated, and possibly rotated.

The issue of triviality was circumvented in early models by requiring universal computation and universal construction. Inspired by biological cells, more recent models (those starting with [13]) have abandoned this requirement by insisting

---

[1]The quiescent state is always a strongly rotation symmetric cell state and is generally included in CA models with weak rotational symmetry.

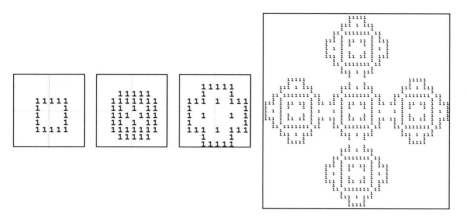

FIGURE 3. Configurations at $t=0$, $t=1$, $t=2$, and $t=22$ for the parity function.

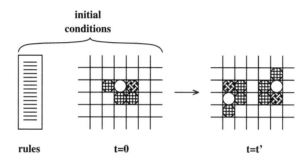

FIGURE 4. Illustration of self-replicating structure.

that an identifiable instruction sequence be treated in a dual fashion: interpreted as instructions (translation), and copied as raw data (transcription). As with unsheathed loops (i.e. loops lacking an outer covering of non-quiescent cells), one can also consider the instruction sequence and the structure itself to be the same, and thus the structure's components directly influence its self-replication process.

2.4. Chronological Summary. A summary of some previous research involving self-replicating structures in cellular space models is shown in Table 2. Most models shown have been 2-D CAs with strong rotational symmetry. In models with weak rotational symmetry, each rotated cell state is counted in the "states per cell" column. The sizes of the self-replicating structures are measured in non-quiescent cells, and are sometimes estimates since some systems were never implemented. The listed models are primarily designs and implementations, though existence proofs of self-replicating structures have appeared (e.g., [**25**]). The models shown include variations of cellular automata, and will be discussed in the next section. Briefly, CT-machines are programmable finite automata with registers, α-Universes are CAs augmented with chemistry-like operators, non-uniform CAs allow cells to have differing rules, and W-machines are Turing machine models that are programmable using high-level instructions.

Year	Model Type	Rot. Symmetry	States per Cell	Neighborhood size(s)	Structure size(s)	Functionality[*]	Refs.
1951	CA	weak	29	5	$> 10^4$	a	[31, 21]
1965	CA	strong	8	5	$> 10^4$	a	[7]
1966	CT-mach.	weak	$\approx 10^{100}$	5	$\approx 10^2$	a	[1]
1973	CA	strong	8	5	$> 10^4$	s	[29]
1976	α-Univ.	strong	5	var.	≈ 60	s	[10]
1984	CA	strong	8	5	86	s	[13]
1989	CA	strong	6	5	12	s	[3]
1993	CA	†	6,8	5,9	5–48	s	[23]
1995	CA	strong	10	9	52	a	[27]
1995	EA	weak	9,13	5	2,3	s	[16]
1995	non-uni. CA	strong	3	9	5	s	[24]
1996	CA/W-mach.	strong	63	5	127	a	[20]
1997	CA	weak	192	9	4,8,...	a	[5]
1997	CA	weak	12	5	4,8	a	[19]

[*] s=self-replication, a=capabilities in addition to self-replication.
† Both strong and weak rotational symmetries were investigated.

TABLE 2. Comparison of some self-replicating structures in cellular space models.

3. Review of Models

The section presents a series of brief reviews of self-replicating structures in cellular space models. Beginning with the pioneering work of von Neumann and continuing to recent models, the trend toward less complex structures is evident. Diagrams showing the space-time iteration of the cellular space are shown for most of the models surveyed here.

3.1. Von Neumann's Model. Among his other interests in the late 1940s, John von Neumann wanted to gain a deeper understanding into the nature of complex systems. He was keenly interested in how such such systems formed and evolved from collections of numerous simple components [30, 31]. This interest led him to investigate machines that could construct other machines, the so-called universal constructors. Within the set of universal constructors are a special subset of machines – the self-replicating machines. Partly from his interest in biological organisms, he devoted much time and energy to the study of self-replicating machines. His seminal work in this area formed the cornerstone of what is known today as artificial life – synthesis-based approaches to theoretical biologies.

The logical design of his self-replicating automaton consisted of a 29-state, 5-neighbor, weakly rotation symmetric CA, consisting of many millions of cells. An overview of this machine is seen in Figure 5, where the four main areas of the machine are identified: tape, tape control section, construction control section, and constructing arm. The tape contained the description of the desired machine to construct. The tape control area read and interpreted the tape as well as transferred excitation signals. The construction control area extended and sent signals to the construction arm.

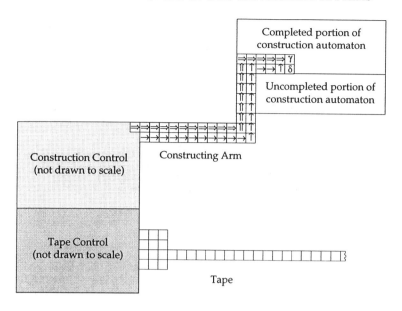

FIGURE 5. Overview of von Neumann's self-replicating automaton (adapted from [2]).

After supplying a correctly-programmed tape to the machine, the sequence of steps needed to have the machine self-replicate were as follows: *i*) reading and interpreting the input tape, *ii*) constructing new cells in the quiescent area, *iii*) "rewinding" the tape, then copying it, *iv*) attaching tape copy to newly constructed portion, *v*) signaling the newly constructed portion that construction had completed, and *vi*) retracting the construction arm.

3.2. Codd's Model. E. F. Codd introduced a simpler universal constructor embedded in an 8-state, 5-neighbor, 2-D strongly rotation symmetric cellular automata, consisting of 100,000,000 cells [7]. A simplification to approximately 95,000 cells appeared later [8]. It shared behavioral similarities to von Neumann's model, but with reduced complexity. The design was influenced by neurophysiology of animals, and one of the notable features was the inclusion of sheathed signal paths.

3.3. Vitányi's Model. The model of Vitányi [29] was an example of a sexually-reproducing cellular automaton. This model employed an 8-state, 5-neighbor cellular space and requires tens of thousands of cells for the two structures. It was argued that in transitioning from asexual to sexual reproduction, a change was needed in the number and structure of instruction tapes. The model specified **M**-type (male) and **F**-type (female) automata, each containing two, nearly identical instruction tapes. Although the automata were quite complex, the model showed that sexual reproduction of automata is possible, and that the recombination process was somewhat similar to that of nature.

3.4. Langton's Self-replicating Loop. The relatively recent resurgence in modeling self-replicating structures is mainly due to studies conducted by Christopher Langton. By recognizing that computation universality was not required to obviate triviality, he was able to devise a vastly simpler self-replicating structure

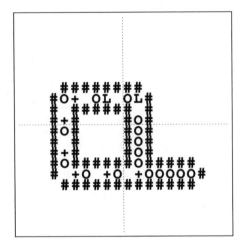

FIGURE 6. Initial configuration of Langton's self-replicating loop. An identifiable instruction sequence ++++++LL is readily seen embedded in the core of 0 states within the sheath.

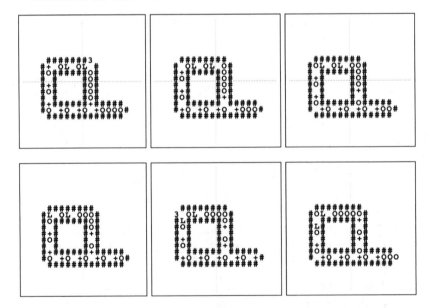

FIGURE 7. Time-steps 1 through 6 for Langton's self-replicating loop. The instruction sequence circulates counterclockwise on successive steps.

in cellular automata [13]. Using concepts from Codd's work, he derived an 8-state, 86-cell sheathed loop that requires 108 replication rules, orders of magnitude simpler than previous models. The initial state of the loop is depicted in Figure 6, and the first six time steps are shown in Figure 7. Figure 8 shows time step 151 where the first replicant has appeared.

3.5. Byl's Model. Byl made further refinements and derived a six state, twelve cell self-replicating structure that required 57 replication rules and had a

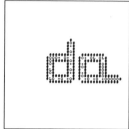

FIGURE 8. Time-step 151 shows the first replicant.

FIGURE 9. Initial configuration of Byl's self-replicating loop.

single sheath [3]. Figure 9 shows the initial configuration of the loop. Figure 10 shows the first 24 time steps of the loop, and Figure 11 shows the first replicant produced at time step 25.

3.6. Reggia's Self-replicating Loops. Further simplification of self-replicating loops was found by deriving unsheathed loops, and varying symmetry conditions [23]. This study verified that both strong and weak rotational symmetries can yield simple self-replicating structures. The smallest structures found were a 6-state, 5-neighbor, 5-cell unsheathed loop under strong rotational symmetry, and an 8-state, 5-neighbor, 6-cell unsheathed loop under weak rotational symmetry. Figure 12 shows the first ten time steps for a structure with a six component unsheathed loop embedded in an 8-state, 5-neighbor CA space with weak rotational symmetry. Figure 13 shows the colony that forms at time step 84.

3.7. Tempesti's Model. In [27] a 6-state, 9-neighbor, 52-cell self-replicating structure is reported that is augmented with additional construction and computational capabilities. It is similar to Langton's self-replicating loop, except in the following ways. First, it has the ability to execute programs in offspring structures. Second, it uses a single interior sheath (instead of a double sheath), which is constructed prior to the signal being sent out. Third, instead of parent loops becoming quiescent, they remain active and are capable of program execution. Fourth, the construction arm extends in four directions simultaneously, as opposed to a single direction. Figure 14 illustrates the structure of the loops at two points in time.

3.8. Arbib's CT-machine. Arbib [1] noticed that the large degree of complexity of von Neumann's and Codd's self-replicating automata could be greatly reduced if the fundamental components were more complex. He developed a model in which automata are analogous to biological cells, as opposed to molecules. Thus, his

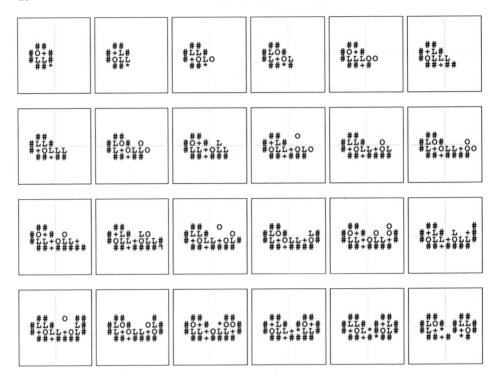

FIGURE 10. Time steps 1 through 24 of Byl's self-replicating loop.

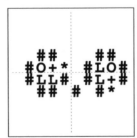

FIGURE 11. First replicant produced after 25 time steps.

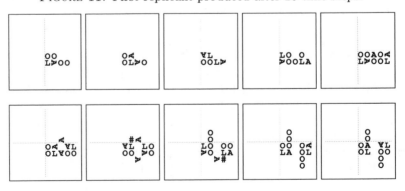

FIGURE 12. Time-steps 0 through 9 for an unsheathed loop structure [23]. The number of replication rules for this structure is 58.

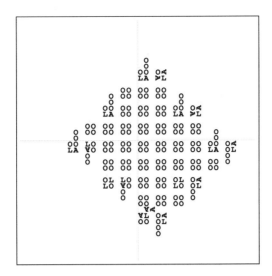

FIGURE 13. The colony that forms later (time step 84) in the development of the unsheathed loop structure from Figure 12.

```
                                                              1
                                                              1
            1                            1                    1
         dddd2ddd                     1dd2ddddd            dddd2ddd
        1d111111d                     d111111d            1d111111d
        d1     1d                     d1     12            d1     12
        21     1d                     d1     1d            d1     1d
        d1     12                     d1     1d            d1     1d
        d1     1d                     21     1d            21     1d
        d111111d1                     d111111d            d111111d111
        ddd2dddd                      ddddd2dd1           ddddd3dd
        1                             1                   1


            time = 0                        time = 121
```

FIGURE 14. Time steps 0 and 121 of Tempesti's loop. State d represents the data state.

automata are very complex, and his description and proofs regarding self-replicating functionality are much shorter than von Neumann's and Codd's.

His automata had approximately 10^{100} states and were capable of both universal computation and construction. The automata are embedded in a 2-D cellular space model called Constructing Turing machines, or CT-machines [**28**]. Each cell in this space (Figure 15) contains a finite-state automata that execute short 22-instruction programs (Figure 16). Instructions consist of actions such as weld and move, and internal control constructs such as if/then and goto. Self-replication occurs when individual CT-machines copy their instructions into empty cells. Composite structures consisting of multiple CT-machines are able to move as one unit

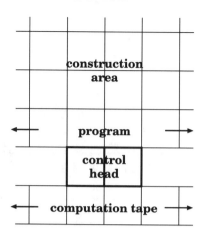

FIGURE 15. Overview of embedded automata in the CT-Machine model.

since individual automata can be welded to each other. Each of the four components is constructed out of identical automata, each programmed specifically for the appropriate function W denotes weld positions, BR denotes bit register, module is programmed using instructions such as weld, emit, move, goto. The machine operates as follows. Since cells can be "welded", a tape can be formed. The control head can read and write the computation tape in the same manner as a Turing machine. The construction area is initially quiescent. Program cells can only write into the construction area, and these write operations are equivalent to the placing of new components.

3.9. Holland's Model. In the mid 1970s, John Holland developed a theoretical framework for the spontaneous emergence of a class of artificial self-replicating systems [10]. Holland defines a set of model "universes" containing abstract counterparts to rudimentary chemical and kinetic mechanisms such as bonding and movement. He wanted to loosely model natural chemical processes (diffusion, activation) acting on structures composed of elements (nucleotides, amino acids) to show that even with random agitations, the tendency of such a system would not be sustained randomness, but rather, life "in the sense of self-replicating systems undergoing heritable adaptations."

The α-Universe cellular space model represents cell states as elements that are logical abstractions of physical entities (e.g. atoms) and obey the conservation of mass. Figure 17 shows an example of part of an α-Universe. Elements are the fundamental units with codons encoding the elements as seen in the illustration.

Interactions among the elements are strictly local as in CA, but some are localized to aggregate structures (strings of bonded elements). Elements behave as automata during the first of three "phases" of each discrete time-step. During the second and third phases, they are acted upon by the four operators: bonding, movement, copy, and decode. As an example, the "copy" operator becomes activated if the sequence $-0:e_1 e_2 \cdots e_l-$ forms (e_i being one of the three elements), and it would cause elements to be reshuffled so that a codon-encoded copy of the string $e_1 e_2 \cdots e_l$ would be assembled.

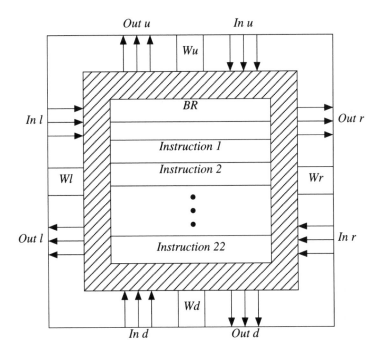

FIGURE 16. Automaton in Arbib's CT-Machine model.

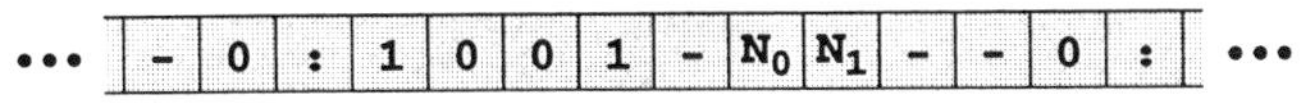

FIGURE 17. An example of a few cells from an α-Universe

Holland parameterizes important aspects of the α-Universes and then uses these to derive formulas that predict the expected time required for emergence of a self-replicating system. Substituting reasonable values into his derivations, a waiting time of 1.4×10^{43} time-steps is computed (no emergence). Relaxing the requirement from *fully* self-replicating to *partially* self-replicating, a waiting period of 4.4×10^8 time-steps (4.4×10^8 seconds is about 14 years) is obtained. Since this is a reasonable amount, it lends credence to spontaneous emergence of self-replicating structures in general, given that Holland's model and derivations are accurate. In [**18**], an empirical investigation claims that some of the conjectures were flawed. Regardless of whether the original analysis is valid, it remains one of the only studies of its kind reported to date and raises important theoretical questions regarding emergence of self-replicating structures.

3.10. Sipper's Model. Sipper describes a self-replicating loop motivated by Langton's work [**24**]. The cellular space model is a modified cellular automata

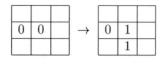

FIGURE 18. Example transition rule.

```
                                                                    1
        11        11        11        11        11      11        11
        111       1110      11100     11101     11101   11011     11 11
                                      1         11      11        11
                                                        1         1
                                                                  0


        t=0       t=1       t=2       t=3       t=4     t=5       t=6
```

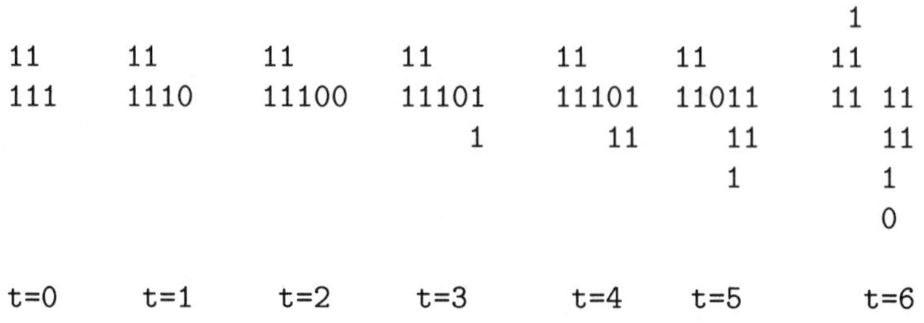

FIGURE 19. Self-replicating structure comprised of 5-cells that uses 10 transition rules in a 2-state, 9-neighbor non-uniform CA model.

whereby: the space iterates in discrete time with cells updated in a local, synchronous manner, but unlike a CA, a given cell can change a neighboring cell's state (Figure 18 shows an example transition rule). Also, a cell can copy its rule into a neighboring cell (non-uniform CA). Figure 19 shows the five component structure in its self-replication process.

3.11. Perrier's Self-replicating Loop. Perrier reports a 63-state, 5-neighbor, 127-cell self-replicating structure, exhibiting universal computation[20] (see Figure 20). Universal computation is achieved by using Turing machine model called the W-Machine which is programmed using a small instruction set. Complexity is reduced by eliminating requirement of construction universality. The loop structure self-replicates in the same manner as Langton's self-replicating loop. Program and data tapes are copied using transmitted signals. After a daughter structure is produced, it can execute a W-Machine program.

3.12. Emergent Self-Replicating Structures. Previous self-replicating structures have always been initialized with a pre-defined structure. Chou and Reggia [5] investigates whether there exists a CA transition rule that can promote the emergence of self-replicating structures from a *randomly* initialized CA space. Using a new cellular automata programming language and development environment called Trend, CA transition rules were found that: *i*) support replication of different-sized structures, *ii*) show growth of small structures into larger ones, *iii*) allow interactions between structures, and *iv*) are robust: independent of space size and initial component density. Figure 21 shows some of the emergent self-replicating loops that emerged.

3.13. Self-Replicating Loops: Problem Solving and Artificial Selection. Recent work on models that incorporate problem-solving capabilities into self-replicating loops has yielded loops that can solve satisfiability (SAT) problems. Previous models incorporated a fixed "program" that is copied unchanged to replicants. Chou and Reggia [6] demonstrate solutions to the SAT problem

```
........
.70170170.
.1......1.
.1.    .7.
.1.    .0.
.1.    .1.
.1.    .7.
.0......0....
.410410710711.
.A...........
.P.    .D.
.P.    .D.
.P.    .D.
.P.    .D.
.P.     .
.P.
.P.
 .
```

FIGURE 20. Perrier's self-replicating loop. The structure consists of three parts: loop, program, and data: D represents a data cell, P represents a program cell, and A represents the position of the program.

in which replicants receive partial solutions that are modified during replication, and artificial selection: promising solutions proliferate, failed solutions are lost. The environment selects satisfied clauses by using "monitor" cells which destroy unsatisfied loop fragments. Figure 22 depicts how the SAT problem predicate $Q = (\neg x_1 \lor x_3) \land (x_1 \lor \neg x_2) \land (x_2 \lor \neg x_3) \land (x_4 \lor x_4) \land (\neg x_4 \lor \neg x_5) \land (x_5 \lor \neg x_6)$ is solved using 4 by 4 self-replicating loops.

3.14. Automatic Discovery of Self-replicating Structures. The question of automatically discovering self-replicating structures is examined in [16, 17]. In all such past models, the underlying transition rules have been manually designed, a process that is very difficult and time-consuming, and is prone to subjective biases of the implementor. This research introduced the use of genetic algorithms to discover automata rules that govern emergent self-replicating processes. Identification of effective performance measures (fitness functions) for self-replicating structures was a key challenge in this problem. A genetic algorithm using multiobjective fitness criteria was applied to automate rule discovery. The results show that novel self-replication processes were uncovered by the genetic algorithm. For example, some of our structures both rotate and move during self-replication, and some leave around unused components (debris) which promote the formation of new structures. Such behaviors, which have not been used or considered in past manually-designed self-replicating structures, are especially interesting, suggesting that evolutionary computation can discover novel design concepts of general value. Figure 23 shows an examples of an automatically discovered structure.

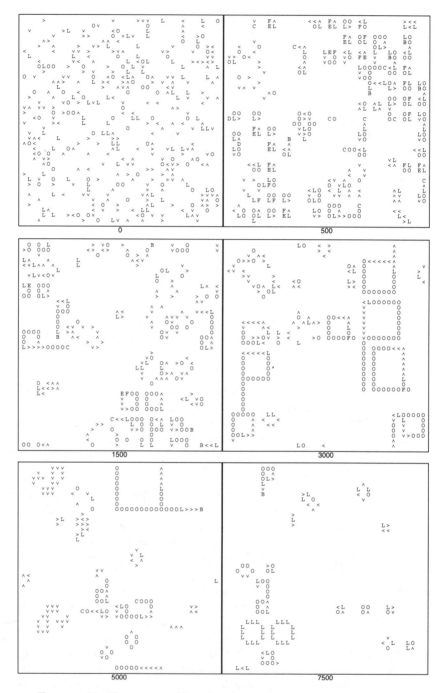

FIGURE 21. Emergent self-replicating loops: at $t = 500$ 2×2 and 3×3 can be seen, at $t = 1500$ 4×4 loops, at $t = 3000$ 8×8 loops, and at $t = 5000$ a single 10×10 loop can be seen. At $t = 7500$ large loops have been replaced by smaller ones. Reprinted from "Emergence of Self-Replicating Structures in a Cellular Automata Space" by H. H. Chou and J. A. Reggia in *Physica D*, (110) 3–4, pp. 252–276, 1997, by permission of the publisher Elsevier Science.

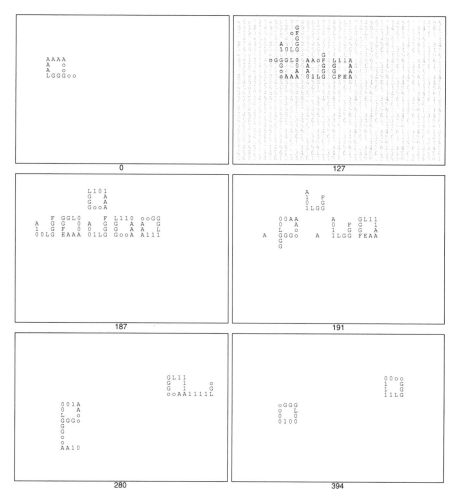

FIGURE 22. Self-replicating loops during SAT problem solving. The initial 4 by 4 loop contains explored bits AAAAAA. The frame at $t = 127$ shows the population of "monitor" cells which function to remove unsatisfied clause bits. New loops are generated and tested by the monitor cells (not shown in last four frames). The two solutions, 000100 and 111100 are found at $t = 394$. Reprinted from "Problem Solving During Artificial Selection of Self-Replicating Loops" by H. H. Chou and J. A. Reggia in *Physica D*, (115) 3–4, pp. 293–312, 1998, by permission of the publisher Elsevier Science.

4. Discussion

Many of the models of self-replicating structures appearing in the literature have been described in the preceding sections. From the early models to the present day, the progressive simplification of self-replicating structures in cellular automata is apparent. This was accomplished first by relaxing the requirements of construction and computation universality, and later by reduction of structure size. We've seen how self-replicating structures have been constructed in cellular space models other than cellular automata, and how automatic discovery methods have been employed with respect to searching for self-replicating structures.

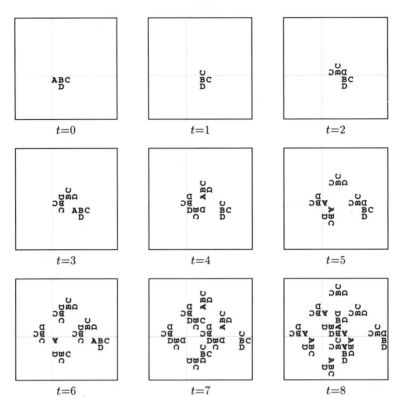

FIGURE 23. Four-component self-replicating structure. The seed structure moves towards the right on successive time steps and produces two replicants: the first is seen at $t=4$ and then again at $t=7$ along with the second replicant (upper right quadrant of each respective frame). These replicants are rotated 90° counterclockwise and proceed upward. During the production of the first replicant, debris forms (near coordinate system origin of $t=3$ and $t=4$) and coalesces into two structures seen at $t=5$, lower left. One structure moves downward and attempts to self-replicate but due to crowding, is unable. The other moves to the left and produces its first replicant at $t=8$ (lower left quadrant).

There are many directions for further research in this area. Investigating minimal structure size in cellular automata structures, the effect of varied neighborhoods (size and shape) and varied seed structures are logical extensions of some of the previous work. Larger questions regarding the choice of cellular space models, for example, stochastic automata, would appear worthwhile. In the realm of automatic discovery, investigation of other search techniques would be of interest since the fitness landscapes in these problems are poorly understood. Another area is biochemical simulation: a few promising studies have appeared in which modified cellular automata models are used to mimic biochemical interactions and simulate template-directed oligonucleotide replication.

References

[1] M. A. Arbib, "Simple Self-Reproducing Universal Automata," *Information and Control*, vol. 9, pp. 177–189, 1966.

[2] A. Burks, Ed., *Essays on Cellular Automata*, Urbana, IL: Univ. of Illinois Press, 1970.

[3] J. Byl, "Self-Reproduction in Small Cellular Automata," *Physica D*, vol. 34, pp. 295–299, 1989.

[4] R. A. Brooks and P. Maes, Eds., *Artificial Life IV*, Proceedings of the Fourth International Workshop on the Synthesis and Simulation of Living Systems, Cambridge, MA: MIT Press, 1994.

[5] H. H. Chou, J. A. Reggia, "Emergence of Self-Replicating Structures in a Cellular Automata Space," *Physica D*, vol. 110, pp. 252–276, 1997.

[6] H. H. Chou, J. A. Reggia, "Problem Solving During Artificial Selection of Self-Replicating Loops," *Physica D*, vol. 115, pp. 293–312, 1998.

[7] E. F. Codd, *Cellular Automata*, New York: Academic Press, 1968.

[8] J. Devore, R. Hightower, "The Devore Variation of the Codd Self-Replicating Computer," Third Workshop on Artificial Life, Santa Fe, New Mexico, draft November 30, 1992.

[9] K. E. Drexler, "Biological and Nanomechanical Systems: Contrasts in Evolutionary Capacity," in [14], pp. 501–519, 1989.

[10] J. H. Holland, "Studies of the Spontaneous Emergence of Self-Replicating Systems Using Cellular Automata and Formal Grammars," in *Automata, Languages, Development*, A. Lindenmayer and G. Rozenberg, Eds., pp. 385–404, 1976.

[11] J.-I. Hong, Q. Feng, V. Rotello, and J. Rebek, Jr., "Competition, Cooperation, and Mutation: Improving a Synthetic Replicator by Light Irradiation," *Science*, vol. 255, pp. 848–850, 1992.

[12] J. O. Kephart, "A Biologically Inspired Immune System for Computers," in [4], pp. 130–139, 1994.

[13] C. G. Langton, "Self-Reproduction in Cellular Automata," *Physica D*, vol. 10, pp. 135–144, 1984.

[14] C. G. Langton, Ed., *Artificial Life*, Santa Fe Institute Studies in the Sciences of Complexity, vol. VI, Reading, MA: Addison-Wesley, 1988.

[15] C. G. Langton, C. Taylor, J. D. Farmer, and S. Rasmussen, Eds., *Artificial Life II*, Santa Fe Institute Studies in the Sciences of Complexity, vol. X, Reading, MA: Addison-Wesley, 1991.

[16] J. D. Lohn and J. A. Reggia, "Discovery of Self-Replicating Structures using a Genetic Algorithm," *1995 IEEE International Conference on Evolutionary Computation*, Piscataway, NJ: IEEE Press, pp. 678–683, 1995.

[17] J. D. Lohn and J. A. Reggia, "Automatic Discovery of Self-Replicating Structures in Cellular Automata," *IEEE Transactions on Evolutionary Computation*, vol. 1, no. 3, pp. 165–178, 1997.

[18] B. McMullin, "The Holland α-Universes Revisited." In *Toward a Practice of Autonomous Systems: Proceedings of the First European Conference on Artificial Life*, F. Varela and P. Bourgine (eds), MIT Press, pp. 317–326, 1992.

[19] K. Morita, K. Imai, "A Simple Self-Reproducing Cellular Automaton with Shape-Encoding Mechansim," in *Artificial Life V*, C. G. Langton and K. Shimohara, Eds., Cambridge, MA: MIT Press, pp. 489–496, 1997.

[20] J.-Y. Perrier, M. Sipper, J. Zahnd, "Toward a Viable Self-Reproducing Universal Computer," *Physica D*, vol. 97, pp. 335–352, 1996.

[21] U. Pesavento, "An Implementation of von Neumann's Self-Reproducing Machine," *Artificial Life*, vol. 2 no. 4, pp. 337–354, 1995.

[22] T. S. Ray, "Evolution, Ecology and Optimization of Digital Organisms," *Santa Fe Institute Working Paper 92-08-042*, 1992.

[23] J. A. Reggia, S. Armentrout, H. H. Chou, and Y. Peng, "Simple Systems That Exhibit Self-Directed Replication," *Science*, vol. 259, pp. 1282–1288, 1993.

[24] M. Sipper, "Studying Artificial Life Using a Simple, General Cellular Model," *Artificial Life*, vol. 2, no. 1, pp. 1–35, 1995.

[25] A. R. Smith, "Simple Nontrivial Self-Reproducing Machines," in [15], pp. 709–725, 1991.

[26] A. H. Taub, *John von Neumann: Collected Works. Volume V: Design of Computer, Theory of Automata and Numerical Analysis*, Oxford: Pergamon Press, 1961.

[27] G. Tempesti, "A New Self-Reproducing Cellular Automaton Capable of Construction and Computation," in *ECAL95: Proceedings of the Third European Conference on Artificial*

Life, F. Moran, A. Moreno, J. J. Morelo, and P. Chacon, Eds., Heidelberg: Springer-Verlag, pp. 555–563, 1995.

[28] J. W. Thatcher, "Universality in the von Neumann Cellular Model," in [**2**], pp. 132–186, 1970.

[29] P. M. B. Vitányi, "Sexually Reproducing Cellular Automata," *Mathematical Biosciences,* vol. 18, pp. 23–54, 1973.

[30] J. von Neumann, "The General and Logical Theory of Automata," in [**26**], pp. 288–328, 1951.

[31] J. von Neumann, *Theory of Self-Reproducing Automata,* A. Burks, Ed., Urbana, IL: University of Illinois Press, 1966.

[32] S. Wolfram, *Cellular Automata and Complexity,* Reading, MA: Addison-Wesley, 1994.

CAELUM RESEARCH CORPORATION, NASA AMES RESEARCH CENTER, COMPUTATIONAL SCIENCES DIVISION, MS 269/1, MOFFETT FIELD, CA, 94035

E-mail address: jlohn@ptolemy.arc.nasa.gov

Lectures on Mathematics in the Life Sciences
Volume **26**, 1999

On the Evolution of Eukaryotes: Computational Models of Symbiogenesis and Multicellularity

Larry Bull

ABSTRACT. In this paper versions of the abstract NKC model are used to examine the conditions under which two significant evolutionary phenomena - symbiogenesis and multicellularity - are likely to occur and why. Symbiogenesis, taken here to refer to the formation of hereditary endosymbioses, is found to be positively correlated with the mutual interdependence between (simulated) organisms. This may be explained by the fact that symbiogenesis reduces the amount of fitness landscape oscillation experienced by coevolving partners. This paper then examines the initial conditions for the emergence of multicellular organisms. Experimental results show that multicellularity without differentiation appears selectively neutral in comparison to equivalent unicellularity, but that differentiation to soma proves beneficial as the amount of epistasis in the fitness landscape increases. This is explained by considering mutations in the generation of daughter cells and their subsequent effect on the propagule's fitness. This may be interpreted as a simple example of the Baldwin effect.

1. Introduction

In this paper the conditions under which symbiogenesis and multicellularity emerge are investigated. Symbiogenesis is the term used to describe the process under which, as the relationship between symbionts evolves in the direction of increasing dependency, "a new formation at the level of the organism arises - a complex form having the attributes of an integrated morphophysiological entity" [15, p.5]. In a more restricted sense symbiogenesis refers to the formation of hereditary endosymbioses, the symbiotic associations in which partners exist within a host partner and are passed to its offspring. In this paper a simple form of hereditary endosymbiosis is examined and its evolutionary progress is compared to that of the equivalent association where the partners do not become so closely integrated. A version of Kauffman's [14] genetics-based NKC model, which allows the systematic alteration of various aspects of coevolving environment, is used to show that the effective unification of organisms via this "megamutation" [10] proves beneficial under certain conditions. Results indicate that the successful emergence of an hereditary endosymbiosis may occur only when the partners are highly interdependent. In the second part of the paper further versions of the NKC model are used to suggest the conditions under which simple multicellularity proves beneficial. It is found that multi-

1991 *Mathematics Subject Classification*. Primary 92D15; Secondary 92-08, 92B05

cellularity without differentiation appears to be selectively neutral in comparison to equivalent unicellularity, but that simple differentiation to soma can prove beneficial as the amount of organism epistasis (both within and between cells) is increased.

The paper is arranged as follows: section 2 describes the NKC model of coevolution used throughout and section 3 contains the results of using it to examine symbiogenesis. In section 4 the model is used to examine the emergence of multicellularity, including results from a version of the model which considers aggregates of unicellular organisms. Finally, all findings are discussed in section 5.

2. The NKC Model

Kauffman [14] introduced the NKC model to allow the systematic study of various aspects of multi-species evolution. In the model an individual is represented by a haploid genome of N genes (two possible alleles), each of whose contribution to fitness depends upon K other genes in its genome (epistasis). Thus increasing K, with respect to N, increases the epistatic linkage, increasing the ruggedness of the fitness landscapes by increasing the number of fitness peaks, which increases the steepness of their sides and decreases their typical heights. Each gene is also said to depend upon C genes in the other species with which the individual interacts. The adaptive moves by one agent/species may deform the fitness landscape(s) of its partner(s). Altering C, with respect to N, changes the extent to which adaptive moves by each agent/species deform the landscape(s) of its partner(s). As C increases mean performance drops and the time taken to reach an equilibrium point increases, where the fitness level of the equilibrium decreases.

The model assumes all intergenome (C) and intragenome (K) interactions are so complex that it is only appropriate to assign random values (Gaussian distribution) to their effects on fitness. Therefore for each of the possible K+C interactions, a table of $2^{(K+C+1)}$ fitnesses is created, with all entries in the range 0.0 to 1.0, such that there is one fitness for each combination of traits. The fitness contribution of each gene of a given genome is found from its individual table. These fitnesses are then summed and normalised by N to give the selective fitness of the total genome (the reader is referred to [14] for full details of the model).

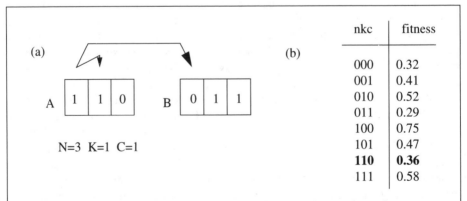

Figure 1: Showing an example NKC model. (a) shows each gene depends on one gene locally and one gene in the other genome. Therefore there are eight possible allele configurations, each of which is assigned a random fitness as shown in (b). Each gene of each genome has such a table created for it. Total fitness of a given genome is the averaged sum of these values.

Kauffman considered populations of one individual (said to represent a converged species) and mutation-based hillclimbing to evolve each species in turn, i.e. each species uses the current context of the others to determine progress. In this paper a generational genetic algorithm (GA)[12] is applied to create a population-based synchronous version of the model; this does not appear to cause the loss of any of the characteristics reported by Kauffman. Relevant observed dynamics are returned to later.

3. Symbiogenesis

"I called this process symbiogenesis, which means: the origin of organisms through the combination and unification of two or many beings, entering into symbiosis" - K.S. Merezhkovsky 1920 [from 15, p.xx].

The most intimate [9] of symbiotic associations is termed endosymbiosis, in which one of the partners, the host, incorporates the other(s) internally. Endosymbiotic associations can be hereditary, wherein the host's endosymbiont(s) pass directly to offspring. The mechanism for this perpetuation ranges from transmission in the egg cytoplasm (transovarial transmission), e.g. in insects, to offspring ingesting the endosymbiont(s) shortly after birth, e.g. when cows lick their calves thereby passing on their rumen ciliates. The formation of an hereditary endosymbiosis can be seen as the logical outcome of the process of symbiogenesis, as defined by Merezhkovsky, and, most significantly, appears to be the way in which some organelles were obtained by eukaryotes [e.g. 17] (it may also have been the way in which chromosomes emerged [e.g. 20]).

In this paper the evolutionary performance of two species (A and B) in two types of symbiotic relationship are compared: symbionts and hereditary endosymbionts (i.e. organisms having undergone symbiogenesis).

Symbiotic organisms: A population (size P) exists for each of the two types of interacting organisms. Individuals are evaluated in pairs, one from each species, and picked consecutively such that individuals are only evaluated once per generation. Note that there is no ordering of populations during reproduction and hence such partnering is random. Each individual is evaluated on its species' NKC function given its current partner.

Hereditary Endosymbionts: A population of hereditary endosymbiotic organisms consists of P individuals carrying the genes of both species A and species B. Species A is arbitrarily said to be the host. Individuals are evaluated by applying the appropriate NKC function to each "part".

3.1. Genetic Algorithm Model. A standard genetic algorithm is applied to Kauffman's NKC model, using fitness proportionate selection ("roulette wheel") and mutation (recombination is not used). A third evolving population also exists to represent the organisms' environment. In this way the effects of increased environmental pressure (interdependence C_e) can be examined by randomly picking an individual from the environmental population (size 2P) to be evaluated with the current symbiotic, or hereditary endosymbiotic, pair. Various values of N have been tried with no significant difference in results being found. N=12 and/or 8 are used throughout this section. All species have the same K value.

It has been stipulated that a symbiotic association must persist for all, or a significant part, of an organism's life span for it to be termed a symbiosis [21]. Further, it has been noted that close symbioses result in an "interspecies supraorganism about as well integrated as parts of an individual organism ... with selection operating on the system as a functional whole" [1, p.718]. To take this into account for the symbiotic organisms modelled here the *combined* fitness of the two partners is used by each organism for selection within its own species. That is, each symbiotic organism is presumed to have spent so much of its lifetime closely associated with its partner that it is only appropriate to also assign it the partner's fitness. The hereditary endosymbiotic organisms also receive a combined fitness. Fitnesses therefore range between 0.0 and 2.0.

It is noted that the important aspect of how the process of symbiogenesis occurs physically, i.e. how endosymbionts are obtained and maintained, are not considered here (e.g. see [8] for a description of how certain organelles may have been obtained).

All experiments consist of running a generational GA over 2000 generations, for 100 trials (10 runs on each of 10 NKC function pairs), for various parameters. Random epistasis patterns (as described in [14, p.244]) are used throughout this paper.

Table 1: Showing the effects of varying the amount of epistasis with regard to which configuration performs best in terms of finding optima (**b**) and mean population performance (**m**) for the two symbionts.

N=8 Cₑ=1

C \ K	1	4	7
b (C=1)	all=	all=	all=
m (C=1)	all=	all=	all=
b (C=3)	all=	all=	all=
m (C=3)	hered	hered	hered
b (C=5)	all=	hered	hered
m (C=5)	hered	hered	hered

N=12 Cₑ=1

C \ K	1	4	7	10
b (C=1)	all=	all=	all=	all=
m (C=1)	all=	all=	all=	all=
b (C=3)	all=	all=	all=	all=
m (C=3)	hered	hered	hered	hered
b (C=5)	all=	hered	hered	hered
m (C=5)	hered	hered	hered	hered

N=8 Cₑ=3

C \ K	1	4	7
b (C=1)	all=	all=	all=
m (C=1)	all=	all=	all=
b (C=3)	all=	all=	all=
m (C=3)	hered	hered	hered
b (C=5)	all=	all=	all=
m (C=5)	hered	hered	hered

N=12 Cₑ=3

C \ K	1	4	7	10
b (C=1)	all=	all=	all=	all=
m (C=1)	all=	all=	all=	all=
b (C=3)	all=	all=	all=	all=
m (C=3)	hered	hered	hered	hered
b (C=5)	all=	all=	all=	X
m (C=5)	hered	hered	hered	X

3.2. Results. Table 1 and Figure 2 show the general result for population size P=50, mutation rate set at probability 0.01 per bit per generation and $C_e=1$, for various K. It can be seen that as the amount of intergenome epistasis (C) increases, for any K, the relative performance of the hereditary endosymbionts (hered) increases. As C increases, the hereditary endosymbionts first do better first in terms of mean fitness (C>2), and then later also in terms of optima found (C>4) - peaking around K=4. That is, the process of symbiogenesis becomes increasingly beneficial as the interdependence between the two symbionts increases. Conversely, when C is low (C<3), the process of symbiogenesis provides no benefits over the symbiotic partners remaining genetically separated; performance is equivalent (all=) under low interdependence. However, under low K (K<3) forming an hereditary endosymbiosis never does better in terms of optima found, but only in terms of mean performance (C>2).

Table 1 also shows a general result wherein the amount of environmental dependence is increased ($C_e=3$). It can be seen (Figure 3) that the benefits of symbiogenesis are lost as C_e increases, with it no longer performing better in terms of optima found and the difference in mean performance decreasing. That is, if the symbionts become increasingly dependent upon other aspects within their environment, forming an hereditary endosymbiosis becomes less likely. Burns' [6] suggestion that some endosymbionts no longer experience environmental factors can be seen as a way in which symbiogenesis may maintain its advantage, i.e. the effects of C_e are lost on endosymbionts.

The consequences of varying the other parameters have also been examined.

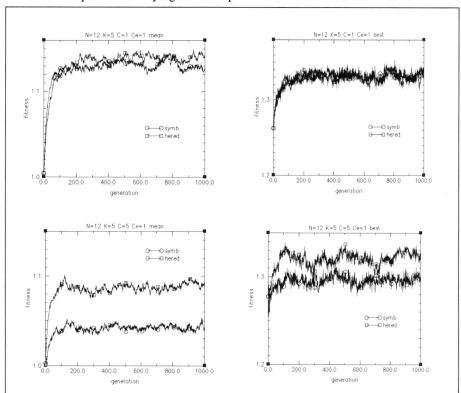

Figure 2: Showing how the hereditary endosymbiotic organisms do better in terms of mean and optimal performance for increasing intergenome epistasis C.

Decreasing the mutation rate (e.g. 0.001) results in the hereditary endosymbionts losing their advantage under all conditions, even high C, in comparison to the separate symbionts (Figure 4). Law [16] notes how obligate symbionts (high C) benefit from having a slower rate of genetic change because they need to stay similar to their parents to keep an association beneficial (also mentioned in [13]). Kauffman [14, p.250] also notes that the optimal mutation rate decreases as C increases. The results (not shown) are similar to those above (0.01) when higher mutation rates (e.g 0.02) are used.

With larger populations (e.g. P=100) it is found that the hereditary endosymbionts do better under increasing interdependence (Figure 5), even for low K (K<3). For smaller P (e.g. P=30) the reverse is true with the separate symbionts doing as well as those having undergone symbiogenesis, particularly in terms of optima found, for higher C (not shown). This is perhaps to be expected, since the hereditary endosymbionts effectively represent a single organism of genome length 2N, and so the larger population size helps the search of their larger fitness space and vice versa for smaller populations.

3.3. Discussion. From these simulations it can be seen that as the interdependence (C) between symbionts increases the benefit of hereditary endosymbioses increases.

Kauffman [14] investigated the effects interdependence has upon separate coevolving partners, finding that:

(i): When the partners are least interdependent (C is low) the time taken for the system as a whole to reach equilibrium is low, fitness during the period of oscillations is close to the equilibrium fitness and the mean oscillating fitness is high.

(ii): When the degree of interdependence is increased the reverse is true with an increase in the time taken for equilibrium to be reached, with fitness during the oscillations being below the equilibrium fitness and where the mean of this oscillating fitness is low.

The results are attributed to the fact that as the amount of interdependence is increased, the effect of each partner on the others' fitness landscape increases; the higher C, the more landscape movement. This represents an extension to Kauffman's single organism NK model, in which he found that:

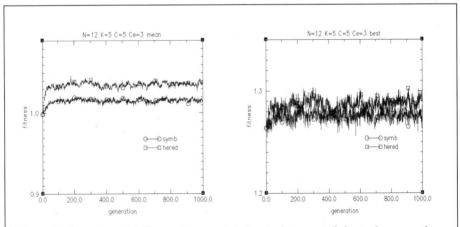

Figure 3: Showing the effects of increasing the environmental dependence on the different organisms.

(iii): Increasing epistasis (K) with respect to N increases the ruggedness of a fitness landscape, but where the height of the local optima decreases.

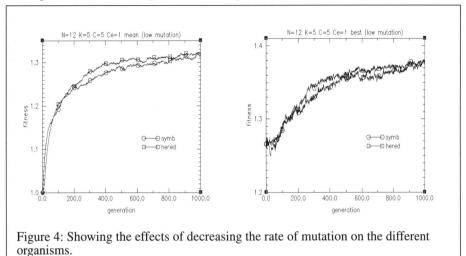

Figure 4: Showing the effects of decreasing the rate of mutation on the different organisms.

Kauffman's findings can be used to suggest underlying reasons for the result above, in that as interdependence increases between symbionts the relative performance of an hereditary endosymbiotic association increases. Hereditary endosymbionts represent two or more genomes being carried together effectively as one genome and as a consequence the partners' intergenome epistasis becomes intragenome epistasis for a larger genome. The effects of intragenome epistasis are very different from intergenome epistasis, as can be seen from *(i - iii)* above. That is, the combined landscape of an hereditary endosymbiosis does not move due to the symbionts' interdependence C - but is more rugged, and larger, than that of the individual partners living separately*(iii)*. There will still be movement from any environmental interdependences (C_e) of course, but now internal epistasis has increased.

Thus, for separate partners, when C is low the amount of possible landscape movement is likewise low. From *(i)* it can be seen that separately coevolving symbionts oscillate with

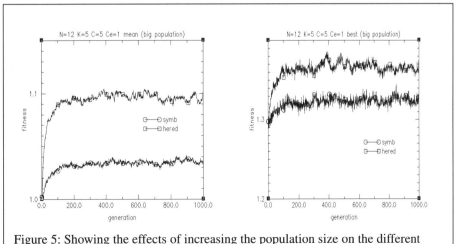

Figure 5: Showing the effects of increasing the population size on the different organisms.

a high mean fitness under these circumstances. Therefore any hereditary endosymbionts must find at least as high a fitness in their larger, more stable, more rugged fitness landscape in order to be selected. They are also hindered by the fact that there is little time for them to search their larger space - even if the fitness level they find is comparable to that of separate individuals - before the separate symbionts reach equilibrium. Thus the evolutionary benefits of symbiogenesis in such cases are low because the separate symbionts can quickly find a high fitness level, and then quickly reach equilibrium; hereditary endosymbioses are unlikely to emerge under such conditions.

When C is high the hereditary endosymbionts' combined landscape will be more rugged, containing many low optima (*iii*). In this case the oscillatory fitness of the coevolving separate symbionts is low (*ii*). Any hereditary endosymbionts will easily find optima in their more stable but rugged landscape and if that peak is comparable to, or better than, the low oscillatory fitness of the separate partners, they will survive into the succeeding generations. The time taken for the separate symbionts to reach equilibrium is longer here (*ii*), which also allows the larger endosymbionts greater searching time.

Kauffman also notes that for any coevolving system "when C is high, increasing the K value of one partner helps *both* coevolving partners" [14, p.249]. Therefore, the process of symbiogenesis can be seen to have a beneficial effect on the significantly coupled ecological partners of the two original symbionts since it results in a higher epistatic organism. That is, symbiogenesis occurs under higher C, where that C becomes K in the supraorganism and so such supraorganisms will typically have a high K value.

A second significant evolutionary phenomenon is now examined.

4. Multicellularity

Approximately 550 million years ago the Cambrian explosion brought forth all the major phyla of multicellular animals. However multicellularity is thought to have evolved up to 200 million years before that and has arisen at least three times - in fungi, plants and animals. In this paper the evolutionary performance of three types of organism are compared: unicellular organisms, multicellular organisms without differentiation and multicellular organisms with simple differentiation. Each organism which survives the selection process is assumed able to divide once. In this way multicellular organisms consisting of two cells are compared to unicellular organisms.

Unicellular organisms: A population (size P) of unicellular organisms consists of P/2 individuals which are the offspring of the previous evolutionary generation (via selection and mutation) and P/2 individuals which are their offspring (with possible mutation); selection works over P individuals to produce P/2 offspring, each of which divides, thus creating a population of P separate individuals. Each individual is evaluated on the given NKC function.

Non-differentiated multicellular organisms: A population of non-differentiated multicellular organisms consists of P/2 offspring from the last generation (via selection and mutation), each of which produces a connected daughter cell (via mutation), i.e. P genomes exist in total. At the end of a generation *all* cells, both propagules and daughters, are able to reproduce; again selection works over P genomes to produce P/2 offspring for the next generation. Both the mother and daughter are evaluated on the given NKC function.

Differentiated multicellular organisms: A population of differentiated multicellular organisms also consists of P/2 offspring from the previous generation, each of which produces a connected daughter cell. At the end of each generation selection works only on the initial P/2 propagules and *not* the daughter cells; the daughter cells are said to have differentiated to soma. Therefore selection works over P/2 individuals to produce P/2 offspring, but again P evaluations occur at each generation since the daughters are also evaluated on the given NKC function.

4.1. Genetic Algorithm Model. A standard GA is again applied to Kauffman's NKC model, using fitness proportionate selection and mutation. The C parameter refers to C_e here, i.e. the organisms' environment, again represented by an extra evolving population of (unicellular) organisms (size 3P). Epistasis between interacting cells will be examined in section 4.3 with additional parameter C_d. Various values of N have been tried with no significant difference in results being found. N=12 and/or 24 are used throughout this section.

Multicellular organisms are formed by a number of binding mechanisms. In higher plants the cells are connected via cytoplasmic bridges and exist within a rigid honeycomb of cellulose chambers. The cells of most animals are bound together by a relatively loose meshwork of large extracellular organic molecules (the extracellular matrix) and by adhesion between their plasma membranes. In all cases the cells exist as part of a larger whole during their lifetime. To take this into account for the multicellular organisms modelled here, an *average* (arithmetic mean) of the two cells' fitnesses is assigned to the reproductive cell(s). That is, it is assumed that the daughter and propagule affect each

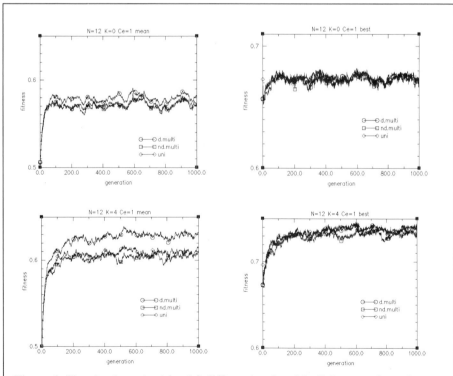

Figure 6: Showing how the (simply) differentiated multicellular organisms do better in terms of mean performance for increasing intragenome epistasis K.

others' fitnesses, and hence the average of their combined fitness is used for selection (this point will be returned to later since it is important). For example, a less fit mother cell producing a fitter daughter would represent a moderately fit organism for selection.

Three important aspects of multicellular origins are not considered here: that of the possible conflict between the mother and daughter (e.g.[7]); details of the mechanism by which the first cells joined (e.g.[22]); and how a mother distinguishes between producing a propagule or daughter cell (e.g.[24]). This paper is concerned with the conditions under which multicellularity could have emerged in a unicellular environment before functional differentiation evolved (most traditional models assume differentiation in the daughter, e.g. improved feeding abilities [24]).

All experiments consist of running a generational GA over 2000 generations, for 100 trials (10 runs on each of 10 NKC functions), for various parameters.

4.2. Results. Table 2 and Figure 6 show the general result for P=100, mutation rate set at 0.01 per bit and C_e=1, for various K. It can be seen that as the amount of intragenome epistasis (K) increases, the differentiated multicellular (d.multi) organisms do better in terms of mean fitness than both the unicellular (uni) and non-differentiated multicellular (nd.multi) organisms. It can also be seen that there is no significant difference between the best and mean performances of the latter two types of organism. That is, the evolution of what is perhaps the first step in the process of becoming a differentiated multicellular organism, that of the propagule staying joined to the daughter cell, appears to be selectively neutral.

Table 2: Showing the effects of varying the amount of epistasis with regard which performs best in terms of finding optima (**b**) and mean population performance (**m**).

K	1	4	7	10
b	all=	all=	all=	all=
m	all=	d.multi	d.multi	d.multi

N=12 C_e=1

K	1	4	7	10
b	all=	all=	all=	all=
m	all=	all=	d.multi	d.multi

N=12 C_e=3

K	1	4	7	10	13	16
b	all=	all=	all=	all=	all=	all=
m	all=	d.multi	d.multi	d.multi	d.multi	d.multi

N=24 C_e=1

K	1	4	7	10	13	16
b	all=	all=	all=	all=	all=	X
m	all=	all=	d.multi	d.multi	d.multi	X

N=24 C_e=3

Table 2 also shows the general result for the same model when the amount of dependence with the environment is increased ($C_e=3$). It can be seen that the amount of intragenome epistasis must increase ($K>4$) before the differentiated multicellular organisms again do better than the others. Kauffman [14, p. 249] also notes that increasing K improves performance under higher C conditions.

The effects of altering the population size and mutation rate have also been examined. Increasing (e.g. P=200) and decreasing (e.g. P=50) the population size does not appear to have any effect on the general result reported above (results not shown). However, the unicellular and non-differentiated organisms often do better when the mutation rate is decreased (e.g. 0.001 per bit), doing as well as the differentiated organisms in terms of mean performance and often better in terms of optima found (Figure 7). Results (not shown) are similar to those above (0.01) for larger rates of mutation (e.g. 0.02 per bit).

A second aspect of the evolution of multicellularity is now considered.

4.3. Multicellularity From the Aggregation of Unicellular Organisms.
Multicellularity is not strictly a eukaryotic phenomenon - it is also seen in prokaryotes through the aggregation of (unicellular) organisms. For example, the rod-shaped Myxobacteria usually live together in loose colonies pooling their digestive enzymes. When food supplies are exhausted or scarce they aggregate into a fruiting body, within which they differentiate, to produce spores that can survive hostile conditions. Unicellular eukaryotes, such as Myxomycota, have similar life-cycles. Eukaryotic green algae range from single-celled organisms (e.g. *Chlamydomonas*) to aggregates of a few cells (e.g. of the genus *Gonium*) to fully multicellular organisms with differentiation (e.g. *Volvox*), prompting the suggestion that multicellularity may have evolved from unicellular aggregates [e.g. 23].

In the previous model the effects on fitness of two cells being joined together during their lifetime was considered by giving the propagule(s) an average of the two "true" fitnesses. However, the NKC model allows inter-genome epistasis to be modelled explicitly (as it was with the parameter C in section 3). It is therefore possible to add a second C parameter to the model which considers the unavoidable epistasis between the two cells (C_d). Once this is done it is no longer possible to compare the two forms of multicellularity with their equivalent unicellular ancestors since the fitness functions will be different -

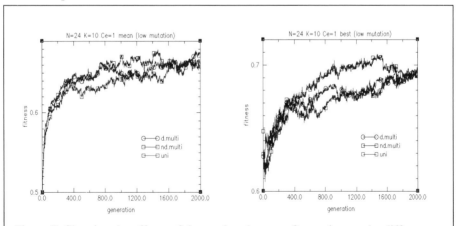

Figure 7: Showing the effects of decreasing the rate of mutation on the different organisms.

unicellular organisms would have tables of $2^{(K+C_e+1)}$ entries whereas multicellular organisms would have tables of size $2^{(K+C_e+C_d+1)}$. Comparing multicellular organisms at various levels of inter-cellular epistasis to equivalent unicellular organisms is possible, however, if the latter are assumed to exist in a colonial form; the interdependence between the individuals in the aggregate is modelled in the same way as the interdependence between the cells of the multicellular organism (C_d). In the work reported here, a unicellular aggregate consists of the initial individual and its offspring [21, p.214].

4.3.1. *Genetic Algorithm Model.* As stated above, the parameter C reflects the effects of others on a given individual's fitness. However, for individuals which spend a significant amount of their life-cycle in very close proximity, the effects of the given individual on the others' fitnesses should also still be considered in evaluating *its* fitness. That is, such a collective of individuals, whether multicellular or an aggregate, should be viewed as a functional whole for selection; an averaged fitness measure of all parts should be assigned to the reproductive individual(s). This reasoning is similar to that used above (section 3.1) in the discussion of close interspecies symbioses.

Here, as above, the reproductive cells of multicellular organisms are given an averaged fitness measure. For the equivalent aggregates of unicellular organisms both possible cases are considered; aggregate models using individual fitnesses and averaged fitnesses are examined. Note that unicellular aggregates with an averaged fitness measure are here equivalent to multicellular organisms without differentiation since mothers are partnered with their daughters.

4.3.2. *Results.* Table 3 and Figure 8 show the general result for P=100, mutation rate set at 0.01 per bit and C_e=1, for various K and C_d, for both types of aggregate. It can again be seen that the differentiated multicellular organisms do better in terms of mean performance under most conditions. Again the difference between the unicellular organisms and non-differentiated organisms is not significant. Increasing C_e also causes the

Table 3: Showing the effects of varying the amount of epistasis with regard which performs best in terms of finding optima (**b**) and mean population performance (**m**) using aggregates of unicells.

C_d \ **K**	1	4	7	10
b 1	all=	all=	all=	all=
m	all=	d.multi	d.multi	d.multi
b 3	all=	all=	all=	all=
m	d	d.multi	d.multi	d.multi
b 5	all=	uni=d	uni=d	all=
m	d	d.multi	d.multi	d.multi

N=12 C_e=1 Aggregate: individual

C_d \ **K**	1	4	7	10
b 1	all=	all=	all=	all=
m	d	d.multi	d.multi	d.multi
b 3	all=	all=	all=	all=
m	d	d.multi	d.multi	d.multi
b 5	all=	all=	all=	all=
m	d	d.multi	d.multi	d.multi

N=12 C_e=1 Aggregate: averaged

advantages of multicellular differentiation to be lost for lower K, as before (not shown).

The effects of altering the population size and mutation rate have also been examined for these two models. Increasing (e.g. P=200) and decreasing (e.g. P=50) the population size does not appear to have any effect on the general result reported above, except that for higher K (K>10) both forms of unicellular aggregate sometimes do better in terms of optima found for bigger P (results not shown). The unicellular and non-differentiated organisms, particularly the unicellular aggregates, again often do better than before (results not shown) when the mutation rate is decreased (e.g. 0.001). Results (not shown) are similar to those above (0.01) for larger rates of mutation (e.g. 0.02).

The use of recombination (not shown) in all of the above multicellularity models does not alter the general result (after [22]).

4.4. Discussion. In the above models it has been found that multicellularity with a simple form of differentiation (to soma) proves beneficial when the fitness landscape is rugged. With the division error rate set low a few selection-generated genomes are evaluated in an environment in which their daughter is slightly different from them. The original genome receives the average of their combined fitnesses; the generated genome gets the average of its "true" fitness and that of a slightly different one. The effect of this is to alter the shape of the underlying fitness landscape for selection since a genome can do better, or worse, than it would on its own. That is, if a given genome can be mutated into a better daughter genome, it must be near a good combination of traits for the function/niche

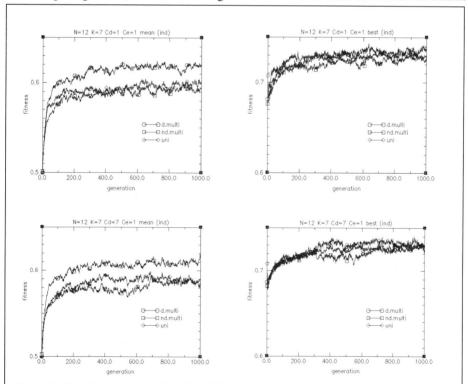

Figure 8: Showing how the (simply) differentiated multicellular organisms do better in terms of mean performance for increasing intergenome epistasis in comparison to non-differentiated organisms and unicellular aggregates (using individual fitnesses).

and so the search is guided toward that region since the original genome operated on by selection gets a higher than expected fitness.

Previously Hinton and Nowlan [11] (and many others since) have shown how life-time learning can achieve the same effect. They showed that a "needle in a haystack" problem can be solved more easily when individuals are given the ability to learn, since the learning can guide evolution to good gene combinations by altering the shape of the fitness landscape; evolution generated genomes receive different (better) fitnesses than they would without learning since they are able to do local (random) search. They describe how learning can turn a difficult problem into a smooth unimodal function because genomes closer to the needle-like optimum are, on average, increasingly able to find it via random learning. This phenomenon is known as the "Baldwin effect" [2]. Mayley [19] has recently applied Hinton and Nowlan's model to a version of the model used here and found that learning becomes increasingly beneficial with increasing K. Therefore the simple form of cell differentiation in multicellular organisms can be seen as an example of the Baldwin effect. The fact that considerably decreasing the mutation rate led to a loss of advantage supports this since less learning is possible in an effectively smaller gene pool; the differentiated organisms often did worse in terms of optima found.

The same fitness altering effect also occurs in both the non-differentiated multicellular case and in the aggregate unicellular case where a combined fitness is used. However, here if a low fitness individual is paired with a good one both receive an average fitness. This means that low fitness genomes look better for selection and will stand a higher chance of producing offspring than appropriate. The fitness landscape altering effect hinders the search process. Conversely, under multicellular differentiation, even when a good genome creates a less fit daughter, the genome which goes back into the population for selection is better than its fitness implies (although the search may be hindered/delayed somewhat by this phenomenon).

Therefore results here indicate that the evolution of differentiated multicellularity appears no more likely to have occurred via unicellular organisms living in aggregates (section 4.3) than via those living alone (section 4.1); the Baldwin effect is caused by the same phenomenon under both conditions and gives the same advantage. However, non-differentiated multicellularity appears selectively neutral and therefore any collective advantage from the new aggregation, such as feeding, in a previously non-aggregating species may have given a selective advantage and started the progress to full multicellularity. The inclusion of costs due to joining to daughters also supports this scenario for the emergence of multicellularity since, if collective advantages through aggregation already exist, the cost of joining would greatly reduce the chances of it giving any net benefit; the difference between unicellularity and non-differentiated multicellularity is no longer neutral when costs are considered. Certainly, multicellularity from aggregation appears to be an evolutionary dead end [22].

5. Conclusions

Symbiogenesis has been the major factor in one of the most important steps in evolutionary history - the evolution of eukaryotes [e.g. 17]. In this paper a tuneable abstract model of multi-species evolution has been used to examine the conditions under which symbiogenesis can occur. It was seen, perhaps not surprisingly, that as the amount of interdependence between the two species increased, the chances of an hereditary endosymbiosis becoming a more advantageous association also increased.

Previously [4] a similar genetic algorithm model was applied to a version of the NKC model to determine the conditions under which symbiogenesis would cause herditary endosymbionts to take over a population of symbionts. It was found that as the amount of dependence between the symbionts increased, i.e. >C, the more dominant the hereditary endosymbionts became.

It is noted that the random partnering strategy used here for symbiont recognition potentially creates larger partner variance than the, often complex, strategies common in nature. The possible effect of reducing partner variance through a more realistic strategy would be to decrease the conditions under which symbiogenesis proves beneficial (see [3] for an effective comparison of recognition strategies). The addition of costs due to the process of symbiogenesis may also cause a reduction in the conditions under which it proves optimal.

In this paper the conditions under which simple multicellularity could have emerged have also been examined. It has been found that multicellularity without differentiation to soma appears selectively neutral in comparison to equivalent unicellularity. However, for fitness landscapes of higher epistasis it has been found that the differentiation of daughter cells to soma proves beneficial in terms of mean performance. The Baldwin effect has been proposed as an explanation of this phenomenon.

The two-celled organisms considered here are more like the early metazoa than plants in that differentiation of the germ-line is assumed to have occurred immediately. For early plant-like multicellular organisms it must have been possible for either the propagule or the daughter to become reproductive. A version of the model presented in section 4 has been used in which either cell randomly becomes the gamete (results not shown). It was found that the advantage of the Baldwin effect is lost and that this form of multicellularity is also selectively neutral; plant-like differentiation did no better or worse than the unicellular and non-differentiated multicellular organisms. This can be explained by considering the case where fitter mothers produce less fit daughters: if the less fit daughter becomes the propagule, it gets a moderate fitness for selection and so its less fit genome has a higher chance of reproducing than expected. Again, any advantage through increased numbers in a previously solitary species, would tip the balance in favour of such multicellularity (section 4.4).

The fact that the multicellular organisms did no better in terms of locating optima may, in part, be due to the characteristics of the NKC model since the previously noted other work on the Baldwin effect using it [19] found similar results. Also, the aggregate model in section 4.3 assumed that mothers were partnered with their daughters during evaluation. If this restriction is lifted such that aggregates are not strictly formed with kin, multicellularity can do better in terms of optima found (not shown - see [5] for an equivalent model).

Finally, the fact that only epistatic/complex organisms appear to benefit from multicellularity is potentially significant. It has been shown how the uptake of hereditary endosymbionts (organelles) reduces environmental instabilities and *increases* the epistasis of the host. This suggests that early (organelle carrying) unicellular eukaryotes were exactly the kind of organisms which would be able to take advantage of the type of differentiated multicellularity modelled here. Indeed, Maynard-Smith and Szathmary [21, p.223] note that since the capacities for gene regulation and cell heredity were already present in (the genetically less complex) bacteria enabling, for example, the distinction between producing propagules and daughters, the evolution of multicellularity may have been limited by a lack of organelles. They suggest that key factors include: the possibilities due to the cytoskeleton; the comparably more efficient photosynthesis/respiration of plasmids/mitochondria; and that there do not appear to have ever been any multicellular archaezoans.

Both models presented in this paper are now being extended to consider more than two participants.

References

1. Allee W C, Emerson A E, Schmidt K P, Park T & Park O, *Principles in Animal Ecology*, Saunders Company, London 1949.
2. Baldwin J M, *A New Factor in Evolution*, American Naturalist **30** (1896), 441-451.
3. Bull L, *Evolutionary Computing in Multi-agent Environments: Partners*, in T Baeck (ed.), *Proceedings of the Seventh International Conference on Genetic Algorithms*, Morgan Kaufmann, San Mateo, 1997, pp. 370-376.
4. Bull L & Fogarty T C, *Artificial Symbiogenesis*, Artificial Life **2** (1996), 269-292.
5. Bull L & Holland O, *Evolutionary Computing in Multi-agent Environments: Eusociality*, in J R Koza, K Deb, M Dorigo, D B Fogel, M Garzon, H Iba & R Riolo (eds.), *Proceedings of the Second Annual Conference on Genetic Programming*, Morgan Kaufmann, San Mateo, 1997, pp. 347-352.
6. Burns T P, *Discussion: Mutualism as Pattern and Process in Ecosystem Organisation*, in H Kawanabe, J E Cohen & K Iwaski (eds.), *Mutualism and Community Organisation*, Oxford University Press, New York, 1993, pp. 239-251.
7. Buss L W, *The Evolution of Individuality*, Princeton University Press, Princeton, 1987.
8. de Duve C, *The Birth of Complex Cells*, Scientific American **274** (1996), 38-45.
9. Ehrman L, *Endosymbiosis*, in D J Futuyma & M Slatkin (eds.), *Coevolution*, Sinauer Associates, Massachusetts, 1983, pp. 128-136.
10. Haynes R H, *Modes of Mutation and Repair in Evolutionary Rhythms*, in L Margulis & R Fester (eds.), *Symbiosis as a Source of Evolutionary Innovation*, MIT Press, Massachusetts, 1992, pp. 40-56.
11. Hinton G E & Nowlan S J, *How Learning Can Guide Evolution*, Complex Systems **1** (1987), 495-502.
12. Holland J H, *Adaptation in Natural and Artificial Systems*, University of Michigan Press, Ann Arbor, 1975.
13. Ikegami T & Kaneko K, *Computer Symbiosis - Emergence of Symbiotic Behavior Through Evolution*, in S Forrest (ed.), *Emergent Computation*, MIT Press, Massachusetts, 1991, pp. 235-243.
14. Kauffman S A, *The Origins of Order: Self-organisation and Selection in Evolution*, Oxford University Press, New York, 1993.
15. Khakhina L N, *Concepts of Symbiogenesis: History of Symbiogenesis as an Evolutionary Mechanism*, Yale University Press, Yale, 1992.
16. Law R, *Evolution in a Mutualistic Environment*, in D H Boucher (ed.), *The Biology of Mutualism; Ecology and Evolution*, Croom-Helm, London, 1985, pp. 145-170.
17. Margulis L, *Origin of Eukaryotic Cells*, Yale University Press, Yale, 1970.
18. Margulis L, *Symbiosis in Cell Evolution*, W H Freeman, New York, 1992.
19. Mayley G, *The Evolutionary Cost of Learning*, in P Maes, M Mataric, J-A Meyer, J Pollack & S W Wilson (eds.), *From Animals to Animats 4*, MIT Press, Massachusetss, 1996, pp. 458-467.
20. Maynard-Smith J & Szathmáry E, *The Origin of Chromosomes I: Selection for Linkage*, Journal of Theoretical Biology **164** (1993), 437-446.
21. Maynard-Smith J & Szathmáry E, *The Major Transitions in Evolution*, W H Freeman, New York, 1995.
22. Szathmáry E, Toy Models for Simple Forms of Multicellularity, *Journal of Theoretical Biology* **169** (194), 125-132.
23. Whittaker R, *New Concepts of Kingdoms of Organisms*, Science **163** (1969), 150-160.
24. Wolpert L, *The Evolution of Development*, Biological Journal of the Linnean Society **39** (1990), 109-124.

FACULTY OF COMPUTER STUDIES & MATHEMATICS, UNIVERSITY OF THE WEST OF ENGLAND, FRENCHAY, BRISTOL BS16 1QY, U.K.

E-mail address: larry@ics.uwe.ac.uk

Lectures on Mathematics in the Life Sciences
Volume **26**, 1999

Cooperation and Conflict in the Evolution of Individuality. III. Transitions in the Unit of Fitness

Richard E. Michod and Denis Roze

ABSTRACT. The evolution of multicellular organisms is the premier example of the integration of lower levels into a single, higher-level individual or unit of fitness. Explaining the transition from single cells to a multicellular organism is a major challenge for evolutionary theory. We provide an explicit genetic framework for understanding this transition in terms of the increase of cooperation among cells within-groups and the regulation of conflict within the cell group—the emerging organism. Cooperation is the fundamental force leading to new levels of organization and selection. While taking fitness away from lower level units (its costs), cooperation increases the fitness of the new higher level unit (its benefits). In this way, cooperation may create new levels of selection and higher levels of fitness. However, the evolution of cooperation sets the stage for conflict, represented here by the increase of deleterious mutants during development. The evolution of a means to regulate this conflict is the first new function at the organism level. The developmental program evolves so as to reduce the opportunity for conflict among cells. An organism is more than a group of cells related by common descent; to exist organisms require adaptations that regulate conflict within. Otherwise, continued improvement of the organism is frustrated by within-organism variation and change during development. The evolution of modifiers of within-organism change are a necessary prerequisite to the emergence of individuality and the continued well being of the organism. Heritability of fitness and individuality at the new level emerge as a result of the evolution of organismal functions that restrict the opportunity for conflict within and ensure cooperation among cells. Conflict leads—through the evolution of developmental adaptations that reduce it—to greater individuality and harmony for the organism.

1. Introduction

1.1. Fitness. Lewontin once remarked that evolution by natural selection should explain "fitness" [29]. In a recent book [39], one of us has taken the approach that to explain fitness we need to understand three things: (i) how fitness originated in the transition from the non-living to the living, (ii) the role of fitness in the mathematical theory of natural selection, and (iii) how new levels of fitness are created during evolutionary transitions to greater levels of complexity. The present paper is concerned with the last question, and in particular how fitness emerges at the level of the

Mathematics Subject Classification. 92D15, 92C15. The authors appreciate the comments of C. Lavigne and C. Nehaniv and the support provided by grants from the NSF (DEB-95277716) and the NIH (GM-55505).

organism out of a group of independently replicating cells. More generally, we are interested in understanding evolutionary transitions to higher levels of organization and complexity. Although, we primarily address the transition from single cells to multicellular organisms, we believe our results are applicable to other evolutionary transitions, such as the transition from replicating genes to cooperating gene networks, from gene networks to the origin of the first cell, from bacteria cells to eukaryotic cells, and the transition from multicellular organisms to societies.

1.2. A scenario for the origin of multicellular life. To help fix ideas, let us consider a scenario for the initial stages of the transition from unicellular to multicellular life (Figure 1).

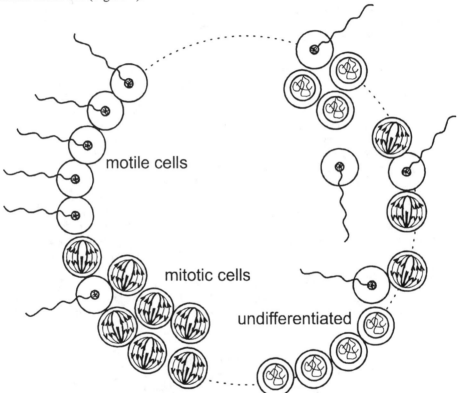

Figure 1. Scenario for the first organisms (groups of cells). Shown in the figure are motile cells with a flagella, non-motile mitotically dividing cells and cells which have yet to differentiate into either motile or mitotic states. Because of the constraint of a single microtubule organizing center per cell, cells cannot be motile and divide at the same time. As explained in the accompanying text, motile cells are an example of cooperating cells and mitotically reproducing cells are an example of defecting, or selfish, cells.

In simple multicellular organisms like Pleodorina in the Volvocales [5] or the sponge Leucosolenia [9], the organism is a hollow sphere with at the top a fixed proportion of cells remaining ciliated and dying at the end of the life cycle; the other cells keeping the capacity to divide. We assume that reproduction and motility are two basic characteristics of the early single celled ancestors to multicellular life, and these single cells were able to differentiate into reproductive and motile states [9,30-32]. Cell development was likely constrained by a single microtubule organizing center per cell, and, consequently, there would have been a trade off between reproduction and motility, with reproductive cells being unable to develop flagella for motility and motile cells being unable to develop mitotic spindles for cell division [9,30,31]. Single cells would switch between these two states according to environmental conditions. Finally, the many advantages of large size (escape from predators being just one possible advantage [6,19,46,47]) might favor single cells coming together to form cell groups. We imagine the appearance of a new mutation perhaps coding for a cell adhesion molecule or the collar structures that hold cells together in a *Proterospongia* colony [27]. It is at this point that our investigations begin.

If and when single cells began forming groups, the capacity to respond to the appropriate environmental inducer and differentiate into a motile state would be costly to the cell, but beneficial for the group (assuming it was advantageous for groups to be able to move). Because having motile cells is beneficial for the group, but motile cells cannot themselves divide, or divide at a lower rate within the group, the capacity for a cell to become motile is a costly form of cooperation, or altruism. Loss of this capacity is then a form of defection, as staying reproductive all the time would be advantageous at the cell level (favored by within-group selection), but disadvantageous at the group level (disfavored by between cell-group selection). We are lead, according to this scenario (and many others), to consider the fate of cooperation and defection in a multi-level selection setting during the initial phases of the transition from unicellular life to multicellular organisms.

1.3. Cooperation and conflict. New evolutionary units begin as cooperative groups of existing units. Cooperation is the primary creative force in the emergence of a new unit of selection, because it trades fitness at a lower level (its costs) for increased fitness at the group level (its benefits). In this way, cooperation can create new levels of fitness (Table 1). Two issues are central to the creation of a new unit of selection— promoting cooperation among the lower level units in the functioning of the group, while at the same time mitigating the inherent tendency of the lower level units to compete with one another through frequency dependent fitness effects.

Cooperation represents the benefit of group living; groups of existing units can behave in new and useful ways. Frequency dependent interactions among evolutionary units are both a source of novelty for the group as well as a

Cell	Level of Selection	
Behavior	Cell	Group (organism)
Defection	(+) replicate faster or survive better	(-) less functional
Cooperation	(-) replicate slowly or survive worse	(+) more functional

Table 1. Effect of cooperation on fitness at cell and organism level. The notation +/− means positive or negative effects on fitness at the cell or organism level.

threat to its collective well-being. Cooperation is usually costly to the fitness of the individuals involved. Defection (that is, non-cooperative behaviors) may reap the benefits of the cooperative acts of others and spread in the population, thereby destroying the very conditions upon which its spread depended in the first place. As a result of the spread of defection, cooperation is lost and so is any hope for the creation of a new higher level. Certain conditions are required to overcome the inherent limits posed by frequency dependent selection to the emergence of new levels of selection:

kinship, population structure, and conflict mediation. Conflict mediation is the process by which lower level change is modulated in favor of the new emerging unit. The definition or usage of certain fundamental terms and concepts used in this paper are given in Table 2.

Term or Concept	Definition and Usage
Self-replication	The capacity to make copies, so that even mistakes can be copied
Individual	A unit of selection satisfying Darwin's three conditions of variation, heritability and self-replication with mechanisms to modulate lower level change
Fisherian Fitness	Per capita rate of increase of a variant
Cooperation	An interaction that possibly decreases the fitness of the individual while increasing the fitness of the group
Frequency Dependent Selection	When fitness depends upon interactions within a group or population
Conflict	Competition among lower level units of selection leading to defection and a disruption of the functioning of the group
Selfish Mutation	Mutations that are deleterious at the organism level and advantageous at the cell level (as opposed to mutations that are deleterious at both levels)
Conflict Mediation	The process by which the fitnesses of lower level units are aligned with the fitness of the group
Fitness Covariance	The covariance between individual fitness and heritable genetic properties (used to study the emergence of individuality)

Table 2. Definitions and usage of terms and concepts used in the paper.

1.4.   Multi-level selection. A recent commentary on multi-level selection theory observes whether there is anything in biology that can't be explained by individual selection acting on organisms, that requires selection acting on groups [41]. Although rhetorical, this remark reflects a view often taken in biology that most interesting questions can be addressed by viewing organisms as the sole unit of selection. But where do organisms come from? From single cells, of course. And what are multicellular organisms but cooperative groups of cells related by common descent. In this paper, we extend a multi-level selection framework recently developed to study the evolutionary transition from single cells to multicellular organisms [36]. We argue that multi-level selection theory is needed to explain the origin of the organism—that very creation which is supposed to deny the usefulness of multi-level selection in evolutionary biology.

Organisms can be thought of as groups of cooperating cells. Selection among cells could destroy this harmony and threaten the individual integrity of the organism. For the organism to emerge as an individual, or unit of selection, ways must be found of regulating the selfish tendencies of cells while at the same time promoting their cooperative interactions.

During the proliferation of cells throughout the course of development, deleterious mutation can lead to the loss of cooperative cell functions (such as the ability to become motile in the scenario above). We represent cell function in terms of a single cooperative strategy. Because deleterious mutation leads to the loss of cell function, mutation may produce defecting cells from cooperative cells. Mutant cells no longer take time and resources to cooperate with other cells and as a result may replicate faster or survive better than cooperating cells. Such mutations are disadvantageous at the organism level, but advantageous at the cell level. Deleterious mutation can also produce completely defective cells with no capacity to replicate or survive. Such mutations are disadvantageous at both the cell level and the level of the group. In previous papers, we considered only defecting mutations, however, in the present paper we consider both kinds of mutations. We also extend the multi-level selection framework developed previously [36] by considering other forms of reproduction involving fragmentation in addition to zygote based reproduction, because we wish to understand the origin of the single cell stage which is almost universal in the life cycles of complex organisms. In addition, we have discovered new equilibria for the model at which linkage disequilibrium is positive and at which the population is polymorphic for conflict mediation; for example, cell groups with and without a germ line may coexist.

2. Multi-level Selection Model

2.1. Overview. The sequence of life cycle events involve the creation (through gamete production or fragmentation or aggregation) of a founding propagule or "offspring" group of cells of size N. These offspring grow and develop into an adult and the adult then produces the offspring of the next generation. An overview of the model life cycle is given in Figure 2. In the case of single cell reproduction considered previously [36,37,39,40], $N = 1$, and sex (fusion and splitting with recombination) may occur among offspring propagules.

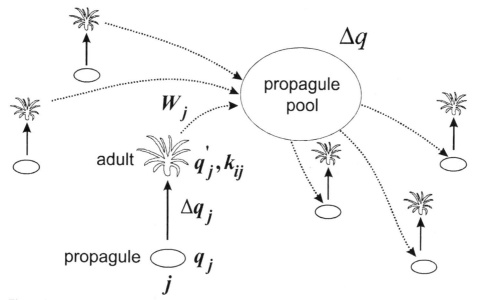

Figure 2. Model life cycle for the origin of organisms. The subscript j refers to the number of *cooperating* cells in a propagule; $j = 0, 1, 2, ...N$, where N is the total number of cells in the offspring propagule group, assumed constant for simplicity. The variable k_{ij} refers to the number of cells of type i (either type C or D) in the adult cell group (that is after development) produced by a propagule of type j. The variables q_j ($q_j = j/N$) and q'_j $\left(q'_j = k_{Cj}/k_j\right)$ refer to the frequency of cooperation before and after development, respectively, while Δq_j is the change in gene frequency at the C/D locus during development. W_j is the fitness of group j, defined as the

expected number of propagules produced by the group, assumed to depend both on size of the adult group after development and its functionality (or level of cooperation among its component cells) represented by parameter β in the models below.

After formation of the offspring, cells proliferate during development to produce the adult form. This proliferation and development is indicated by the vertical arrows in Figure 2. Because of mutation and different rates of replication of different cell types, there will be a change in gene and genotype frequency of cells within-organisms represented by Δq_j in Figure 2. There will also be a change in gene frequency in the population due to differences in fitness between the adult forms. These two components of frequency change, within-organisms and between-organisms give rise to the total change in gene frequency Δq. The number of cooperating and mutant cells of type i in the adult stage of a j offspring is represented by the variables k_{ij} Figure 2. A model to calculate these numbers is given below in Figure 3 and Table 3. The fitness of the adult form, W_j, is the absolute number of offspring produced and this is assumed to depend both upon the number of cells in the adult and how the cells interact. Cooperation among cells increases the fitness of the adult (parameter β in the models below) but non-cooperating cells may replicate faster, survive better, and produce a larger but less functional adult. Organism size is assumed to be indeterminate and to depend on the time available for development as well as rate at which cells divide. Determinate size can also be modeled. The evolution of conflict and cooperation in organisms with a fixed size can be studied by standardizing the frequencies before offspring propagule production much as is done in models of hard and soft selection [52].

2.2. **Within-organism change.** There are two components of gene frequency change: (i) between-organisms within populations, and (ii) between-cells within-organisms. In this subsection we consider a model of within-organism change stemming from mutation during cell division and selection among cells caused by differences in cell replication or cell survival. The basic variables are defined in Table 3 and explained in Figure 3.

As cells proliferate within the developing organism (development time t) , mutations (rate μ) occur leading to loss of tissue function and cooperativity among cells. We consider only mutations that lead to a loss of cooperation (C to D) not its gain (no back mutation from D to C) as this represents a worse case for the evolution of cooperation. This is reasonable for biological reasons, because it is far easier to lose a complex trait like cooperativity among cells than it is to gain it.

Within-organism variation after development is represented by the expected number of cells of different types in the adult form—the k_{Cj} and

μ	mutation rate from C to D per cell division
t	generation time (for development)
c	rate of cell division for cooperating cells
b	effect of mutation on cell replication rate
cb	rate of cell division for mutant cells
s_C	probability of cell survival for C and D
s_D	cells, respectively

$$k_{Cj} = j\left[2s_C(1-\mu)\right]^{ct}$$

$$k_{Dj} = j\sum_{x=1}^{ct} 2\left[2s_C(1-\mu)\right]^{x-1}\mu s_D\left[2s_D\right]^{b(ct-x)}$$

$$+(N-j)\left[2s_C\right]^{bct}$$

$$k_j = k_{Cj} + k_{Dj}$$

Table 3. Mutation and cellular selection model for haploidy. See also Figure 3 for an explanation of how the numbers of different cell types in the adult stage are calculated. Based on model of [42].

k_{Dj} variables given in Figure 2 and defined in Figure 3 and Table 3. Many aspects of the analysis, the various equilibria and their stability for the different reproductive systems, can be obtained analytically without explicitly specifying values for these variables. However, more quantitative analyses require a specific within-organism mutation selection model. The model of within-organism variation and selection presented in this section is a first attempt at such a model. It has the virtue of being comparatively simple, at least in principle, although it becomes complex computationally. There may be other more realistic mutation selection models that could be used to obtain values for the k_{ij} variables, and the model proposed in the last section was set up with this in mind.

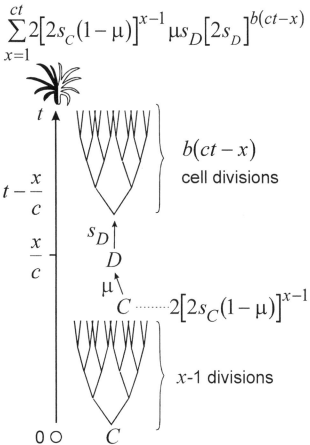

Figure 3. Calculation of the Adult. The figure explains the calculation of the number of defecting cells (D) in the adult stage descended from a single cooperate (C) cell in the offspring propagule (formula given in Table 3). For any single cell generation we assume replication, mutation and then survival. A C cell divides for x-1 divisions and then mutates to D during cell division x. Assume there are N_x cells in generation x. After cell division there are $2N_x$ cells and these survive with probability s_C or s_D, for C and D cells, respectively. The total number of D cells in the adult organism is represented by the color black in the adult form. Consider C cells that have divided, survived and not yet mutated for x-1 divisions and are now in the process of cell division x. The total number of these cells is $2[2s_C (1-\mu)]^{x-1}$. Some of these cells (μ) will mutate for the first time and the resulting mutants will survive to the next cell division with probability s_D and then undergo $b(ct-x)$ more cell divisions. We would like to choose the fitness effect of the mutation at the cell level (b and s_D) from a probability distribution of mutational effects, but for the deterministic treatment given here these mutational effects are fixed parameters of the model and the same for all mutations. The time taken to get x cell divisions is x/c. The time left is $t-x/c$. The number of cell divisions the mutant will undergo is then $cb(t-x/c) = b(ct-x)$. The sum is over all possible cell divisions x. The formulae in the figure simplify to those published previously, if there is no selection on cell survival, $s_C = s_D = 1$ [38]. This approach gives k_{Cj}, k_{Dj} in Table 3 and Table 4 below.

The mutation model given in Table 3 and Figure 3 generalizes to the diploid case following the arguments given elsewhere for a similar model but one without differences in cell survival [37]; however the details are more complex and the steps involved are not presented here for reasons of space. For example, to study the branching process of mutations within cell lineages which start out as a CC cell requires considering four classes of events: (i) CC cells that mutate to CD during cell division x and then mutate to DD during cell division y, (ii) CC cells that mutate to CD during cell division x and remain CD for the rest of development, (iii) CC cells that mutate directly to DD during cell division x and, of course, (iv) CC cells that remain CC for all of development.

The number of cells in an adult depends on the time available for development, the rate of cell division and cell death. Rates of cell division vary widely among tissue types. In some tissues cells stop dividing (brain, muscle, liver) while in other tissues cells continue to divide throughout life (blood, intestine lining). To help fix ideas we can estimate the number of cells that exist after a given development time, t, for cooperating cells as follows. Ignoring mutation, cell death, and the different rates of replication of the different cell types, $t = 40$ would allow 40 cell divisions (assuming $c = 1$) implying around 10^{12} cells in the adult—a number similar in magnitude to the number of cells in an adult human. Including cell death would require a greater number of divisions to get the same number of cells in the adult. Cell death is likely present in most cell lineages. For example, it has been estimated that the number of cell divisions between the zygote and an average human male sperm is around 400 cell divisions [51]. Such a large number of cell divisions is needed to account for cell death. A typical human female egg is separated from the zygote by about 20 cell divisions [51].

2.3. Mutation. Because there are two levels of selection, there are two levels at which to consider the effects of mutation—the cell and the cell-group, or organism. The mutations we consider always lead to the loss of the cooperative function, that is the loss of the benefit β at the group level. This loss may be a pleiotropic consequence of the disruption of other cell functions not directly related to cooperation. In this case, the loss of cooperation may also be deleterious at the cell level, if these functions are important for replication and survival of the cell. We refer to such mutations as uniformly deleterious. According to the relative magnitude of the deleterious effects at the two levels (β and b in the simplest model), uniformly deleterious mutations are best eliminated by within-organism (between cell) or between-organism selection (see Figure 5 below and accompanying discussion).

However, the loss of cooperation may not be a pleiotropic consequence of the loss of cell functions related to cell fitness. Indeed, in Figure 1, mutant cells that are defective in the capacity to differentiate into the motile state may replicate faster or survive better than cooperating cells (because of the constraint of a single microtubule organizing center per cell). By freeing energy or time spent on cooperation, the loss of cooperation may lead to an increase of fitness at the cell level. Such selfish mutations are deleterious at the organism level but advantageous at the cell level. Selfish mutations are only eliminated by between-organism selection, because they are favored by within-organism selection at the cell level.

The model is constructed to study both selfish and uniformly deleterious mutations by considering the effects of mutation on the replication of cells within-organisms in terms of the parameters c and b in Figure 3 and Table 3 (or s_C and s_D if effects on cell survival are to be considered). For purposes of discussion, we consider only effects of deleterious mutation on the replication rate of the cell (parameters c and b), but effects of mutation on survival rates can also be studied (parameters s_C and s_D, in Figure 3 and Table 3). The case of altruism given in Figure 1 is modeled by assuming that mutation produces selfish cells that replicate faster than cooperating cells ($c = 1$, $b > 1$). The case of complete altruism occurs when cooperating cells are not able to survive or replicate at all, but here we focus on intermediate cases of altruism. In addition to the case of selfish mutations, we study uniformly deleterious mutations by considering ranges of the b parameter from zero (completely defective mutants) to unity (mutant cells replicate at the same rate as cooperating cells) ($c = 1$, $0 \leq b \leq 1$).

The deleterious mutation rate in the model pertains to the loss of cooperation and perhaps other functions that affect cell replication and/or survival. Although the model considers a single locus affecting cell behavior, there are likely to be many loci that affect tissue function and cooperativity among cells. For this reason, we like to think of this single locus as representing the cumulative effect of deleterious mutation at all loci leading to loss of cooperation. This is not realistic modeling but provides a better understanding of what we would like the model to represent. The mutation rate parameter in the model does not pertain to cells in modern multicellular organisms but rather to cells on the brink of coloniality during the period in which multicellular life first arose. Indeed, an earlier version of the model predicts that one of the ways of coping with within-organism change is to select for modifiers that lower the mutation rate [36,40]. Modifiers are genes whose effect is to change the value of one of the parameters in the model, such as the mutation rate, or in Section 3 the number of cells in a propagule group, or in Section 4 the parameters of within-organism change. Consequently, the mutation rate in modern organisms likely results from the very processes being modeled here. For these reasons, we consider mutation rates that are high when viewed as pertaining to a single locus in modern organisms.

Modern microbes as diverse as viruses, yeast, bacteria, and filamentous fungi, have a genome-wide deleterious mutation rate of 0.003 per cell division even though they have widely different genome sizes and mutation rates per base pair [13]. For independently acting mutations, a mutation rate of 0.003 yields a low value for the mutation load; the average fitness is approximately $e^{-0.003} \approx 0.997$ [22,23]. The genome wide mutation rate in modern multicellular organisms is much higher than in microbes, for example; *Drosophila* has a genome wide rate (expressed on a haploid basis) of approximately 0.5. Of course, the "organism" we have in mind here is nowhere near as sophisticated as *Drosophila*, being on the threshold of the transition from unicellular to multicellular life. The mutation rate in modern microbes of 0.003 has been attained through the evolution of modifiers over billions of years; these modifiers presumably balanced the benefits of reducing the mutation rate with the physiological costs of doing so. For this reason, we think a much higher overall deleterious mutation rate likely held for the unicellular progenitors of multicellular life.

2.4. Fitness. There are interesting issues in how to model fitness in simple proto-organisms. As already mentioned the fitness of the organism, or cell-group, W_j, is simple enough to define; following standard models in population genetics, we take it to be the absolute number of offspring groups produced. Likewise, the fitness of the cell is straight-forward to define in terms of its rate of replication and survival. The interesting issues concern the relationship between fitness at the two levels and the dependence of organism (cell-group) fitness on the fitness properties of component cells. Cooperation is fundamental to the emergence of cell groups and, so, we take organism fitness to depend upon the level of cooperation among its component cells—using terms of the form $1 + \beta q'_j$ in the models below, where β is the benefit of cooperation and q'_j is the frequency of cooperative cells in the adult stage.

Cooperation among cells increases the fitness of the adult (parameter β) but non-cooperating cells may replicate faster, survive better, and as a consequence produce a larger but less functional adult. The question then concerns whether and how to include the effects of organism (cell group) size in organism fitness. If we ignore the contribution of group size to fitness, we may take $W_j = 1 + \beta q'_j$. The effect of organism size in a j-organism (organisms that start out as propagules with j cooperating cells) may be represented by the term k_j/N, where the total number of cells after development is k_j and N is the number of cells in an offspring propagule (Table 3). If we include the contribution of group size, fitness of the organism may then be taken to be the product of organism size and the effects of cooperation or $W_j = \left(1 + \beta q'_j\right) k_j / N$.

We note that if there were no mutation, including the term k_j/N in the expression of fitness would not cause any variation of fitness with N. If there were no mutation, there would be no within-group variation in replication or survival rates and the adult group size, k_j, would be a linear function of its starting size, N, $\left(k_0 = N 2^{bct} \text{ and } k_N = N 2^{ct}\right)$ and so N would cancel in the expression for W_j. For example, considering just the size component of fitness, k_j/N, for $ct = 3$, with $N = 50$ the adult size is $50 \times 2^3 = 400$ and the organism fitness is $400/50 = 8$, and with $N = 1$ the adult size is 8 and the fitness $8/1 = 8$. Consequently, it is not true that just because organism fitness is divided by N, selection will always act to reduce N, so as to increase W_j.

We think it most appropriate to include group size in organism fitness, for simple organisms on the threshold of multicellular life. However, constructing fitness in this way, $W_j = \left(1 + \beta q'_j\right) k_j / N$, underscores the lack of true individuality in these early cell groups, since there is a direct contribution of cell fitness to organism fitness. Even in the case of no interactions between cells, that is no cooperation or defection, if C and D cells had different intrinsic rates of replication or survival, there would be different sizes among cell groups and different fitnesses of these groups according to k_j/N. Yet, these differences in group fitness would have nothing to do with the interactions among the component cells. For individuality to emerge at the cell-group level, fitness at the new level must be decoupled from the fitness of the component cells.

As individuality at the cell group level becomes better defined, we expect organism size to be regulated by factors other than intrinsic cell replication rates. Modifiers affecting propagule size are studied in 3 and those affecting within-organism change and conflict are studied in Section 4. Conflict modifiers regulating organism size are expected to evolve so as to remove one of the advantages of cell defection, that advantage being larger groups (J. Li and R. E. Michod, unpublished results). By forcing all organisms, to attain a constant size, regardless of their proportion of cooperating and defecting cells, determinant size can be seen as an adaptation to regulate conflict. With determinate size, organism fitness would no longer depend on the term k_j / N.

The simple models considered here (Table 3 above or Table 5 below) assume a linear dependence of adult fitness on numbers of cooperating cells in the group, although more complex models could be easily incorporated into the framework considered there. At some point in the evolution of organisms, tissue function and interaction became so integral that the organism could not survive without it. Representing this (zero fitness below a threshold level of cooperation) requires nonlinear fitness models. Multiplicative models and other nonlinear formulations can also be considered in the same framework and can be expected to affect the outcome of the model, much as is the case in the theory of kin selection studied during previous project periods (see, for example, [33,34]).

3. Origin of the Single-Cell Stage of the Life Cycle

3.1. Overview. We wish to understand why the life cycle of most multicellular organisms passes through a single cell stage. This single cell stage is present in a wide variety of plants, animals, and fungi, from simple multicellular organisms in the Volvocales to complex animals such as humans. We follow the terminology proposed by others and use the word "zygote" to refer to the single cell produced during sexual life cycles and the word "stem cell" to refer to the single cell produced during asexual life cycles [15]. Our initial modeling framework involves asexual life cycles and the origin of the stem cell. However, our longer term goal is to study sexual life cycles. To understand why most life cycles involve a single cell stage we must consider alternative forms of reproduction such as fragmentation (as occurs in *Hydra* [8], in colonial choanoflagellates such as *Proterospongia* [27], or in colonial green algae [6]), vegetative reproduction (as occurs in many plants) or aggregation (as occurs in the social Myxobacteria and in the cellular slime molds). As already mentioned, we use the word "propagule" to refer to the offspring of a reproductive process in which a sample (assumed below to be of size N) of cells is taken from the adult to produce the next generation. The way in which the sample is taken affects the level of variation between and within-groups (binomial sampling is assumed below for simplicity).

As discussed in the previous section, organisms are viewed as groups of cooperating cells. As cells divide the group increases in size and propagule groups may fragment off giving rise to the offspring of the next generation. Although, this may happen continuously, for mathematical simplicity a discrete generation approach is taken initially. Organisms (cell groups) are assumed to be composed of cooperating cells (for example motile cells as in Figure 1) and mutant cells. Mutant cells may be completely defective in all functions or defective in the capacity to differentiate into the motile state, as considered in the motivating scenario given in Figure 1. Offspring propagules are denoted by the number of cooperating cells in the propagule group, $j =$ 0, 1, 2, ...N, where N is the number of cells in the offspring propagule group, assumed constant for simplicity. The problem of the evolution of single-cell reproduction is basically the problem of the reduction of N to $N = 1$.

During development deleterious mutations (C to D) occur at rate μ. These mutations may be caused by errors in DNA replication (genetic mutations) or errors in the chromosome marking systems. Because the framework for individuality considered here involves two levels of selection, the cell and the group or whole organism (see Figure 1 and Figure 2), mutation may have deleterious effects at either level (or both). Deleterious mutation may lead to the loss of the benefits of cooperation at the cell group or organism level (for example, loss of motility or the capacity to respond to the motility inducer as in Figure 1), or to the loss of other cell functions related to cell replication and survival (as modeled by effects on the cell replication and survival parameters in Figure 3 and Table 3. The number of cooperating and mutant cells in the adult stage of a j propagule is k_{Cj} and k_{Dj}. A model to calculate these numbers is given in Figure 3 and Table 3 above.

3.2.   Propagule model There are assumed to be two alleles, cooperate C and mutant D. The notation D is for Defect to describe the interesting case of selfish mutations that lead to the loss of cooperation with enhanced rates of cell replication or survival. However, deleterious mutations that disrupt cell function at both the level of the group and the cell are also studied as described in the last section.

We refer to "organisms" in terms of the composition of the offspring with regard to the number of cells that are cooperative at the cooperation locus. Because of within-organism mutation and selection during development, the adult stage may have cells that differ genetically from those in the offspring. The k-variables in Table 4 refer to numbers of cells of different genotypes in the adult stage. There are two forces that may change gene frequency between the zygote and adult stages and determine the values of the k-variables: mutation and cellular selection. Mutation leads to the loss of cooperation, and other aspects of cell function. Mutation increases the variance among cells within-organisms and enhances the scope for selection and conflict among cells. Cellular selection is assumed to depend on differences in cell survival or the rate of cell replication (Table 3, Figure 3). The definition of terms and variables in the haploid model is given in Table 4. Diploidy has also been considered previously in a simpler model

i,j	number of cooperating cells in offspring propagule group $i, j = 0, 1, 2, ...N$
N	number of cells in the offspring propagule group, assumed constant for simplicity
x_i	frequency of propagules with i C cells at time t
f_{ij}	frequency of propagule i produced from the adult form of propagule j
k_{Cj}	number of C cells in the adult stage of a j-propagule
k_{Dj}	number of mutant cells in the adult stage of a j-propagule
k_j	total number of cells in adult stage of j-propagule after development, $k_j = k_{Cj} + k_{Dj}$
W_j	adult fitness (expected number propagules) of a j-propagule taken in Section 3 to be either $$W_j = 1 + \beta q'_j \text{ or } W_j = \left(1 + \beta q'_j\right)\frac{k_j}{N}, \text{ depending on}$$ whether we assume that adult fitness depends on adult size (see subsection 2.4 for discussion)
β	benefit to adult organism of cooperation among cells
q'_j, q_j	freq. in adult and offspring for C gene for propagules of type j
$q', \Delta q$	freq. and change in freq. of C gene in total population

Table 4. Haploid Propagule Model.

[37], but here we focus on the haploid case. With these definitions it is straight forward to write down the new frequency in the next generation of propagules of type i (given in Equation 1),

$$x_i' = \sum_{j=0}^{N} f_{ij} \frac{W_j}{\overline{W}} x_j , \qquad \text{Equation 1}$$

with $\overline{W} = \sum_{i=0}^{N} W_i x_i$.

3.3. Modes of fragmentation. The parameter f_{ij} is the frequency of type i propagules among the offspring of a type j organism. This frequency depends on the fragmentation mode considered. Kondrashov has studied the mutation load (defined as the difference in average fitness with and without mutation) in a population where the individuals produce offspring composed of N cells (as in our model) [26]. The model that he used is similar to ours, but he doesn't include within-organism selection. He considers four possible modes of fragmentation (see his figure 1 [26]).

False mode: the N cells of each offspring are as genetically closed as possible. In fact this case is equivalent to a single-cell-stage, and N has no influence on the mutation load.

Sectorial mode: the N cells come from the same sector; this means that they all come from the same initial cell. One can calculate f_{ij} in the following way. The number of cells in the adult that arise from each of the j initial C cells is

$k_C = 2^{ct}(1-\mu)^{ct} + \dfrac{\mu 2^{bct} - 2^{ct}(1-\mu)^{ct}\mu}{-1 + 2^{b-1} + \mu}$, so the probability of choosing a sector that

arise from a C cell is $\dfrac{jk_C}{jk_C + (N-j)2^{bct}}$. The probability of having an offspring with i

C cells and N-i D cells is for $i > 0$: $f_{ij} = \dfrac{jk_C}{jk_C + (N-j)2^{bct}} \dbinom{N}{i}(q_C)^i(1-q_C)^{N-i}$, with

$q_C = \dfrac{2^{ct}(1-\mu)^{ct}}{k_C}$ being the frequency of C cells among the cells arising from a C

initial.

Random mode: the N cells are taken randomly from the adult. In this case we have Equation 2.

$$f_{ij} = \binom{N}{i}\left(q_j'\right)^i\left(1-q_j'\right)^{N-i} . \qquad \text{Equation 2}$$

Recall, q_j' is the frequency of cooperating cells in the adult form. The frequency of C cells in the offspring and adult stage of a j offspring is $q_j = j/N$ and $q_j' = k_{Cj}/(k_{Cj} + k_{Dj})$, respectively. The k_{Cj} and k_{Dj} variables are calculated Table 3 and Figure 3. To illustrate the model framework, we assume for simplicity Equation 2 that involves random sampling across all cells in the adult form (with replacement). The hypergeometric distribution would be more appropriate if the number of cells in the offspring is big, but it can be approximated by a binomial distribution if the number of cells in the offspring is small compared with the number of cells in the adult k_j (approximately $N < k_j/10$).

Structured mode: each of the N cells come from a different initial cell (case 4). Kondrashov shows that in this case the genetic load is maximal. We will consider the sectorial and random cases in the analysis of the model.

To sum up, the different parameters of the model are represented in Table 3 and Table 4. We note that this model could easily be modified to study the case of groups formed by aggregation of cells (like slime molds). The model is deterministic and it assumes an infinite population with discrete generations. Organisms are haploid and asexual. Each organism begins its development with N cells. Cells can be of two types: C (cooperative) or D (selfish with $b > 1$, or uniformly deleterious with $b < 1$). During development, mutations $C \to D$ occur at a rate μ per cell division. The parameters representing the relative importance of intra- and inter-group selection are b and β. Two modes of fragmentation are considered next (sectorial and random) in addition to spore reproduction, and fitness may depend upon the size of the organism as well as on the frequency of mutants in the organism.

3.4. Mutation load.   Kondrashov calculates the mutation load (difference in average fitness of populations with and without mutation) under vegetative reproduction, using a similar model to ours but without including intraorganismal selection [26]. Otto and Orive extended Kondrashov's model to include intraorganismal selection for uniformly deleterious mutations (they did not study selfish mutations), and show that with moderate intraorganismal selection, the opposite result obtains, that is mutation load decreases when N increases [42]. This indicates that it can be advantageous to produce multicellular propagules if intraorganismal selection can act upon the initial cell variants to eliminate mutations which are deleterious at both levels. In Otto and Orive as in Kondrashov's model, the fitness of the organism doesn't depend on its size (whereas it can in our model as discussed in subsection 2.4). As we will see including organism size in organism fitness modifies the results.

The model with $N+1$ recurrence equations $(x_0, x_1, ... x_N)$ is too complex to be treated analytically. Nevertheless Kondrashov shows that the mutation load at equilibrium $\left(L = 1 - \overline{W}/W_{max} \right)$ can be calculated and is equal to $1 - \lambda$, λ being the first eigenvalue of the matrix $\left[f_{ij} W_j \right]$. The equilibrium distribution of the different types of organisms is given by the first eigenvector of the matrix. He shows that the load increases with N, and is more important with mode 2 of fragmentation than with mode 1, with the mode 3 than with mode 2 and with mode 4 than with mode 3 (1: false mode, 2: sectorial mode, 3: random mode, 4: structured mode, see figure 1 of Kondroshov [26]). This seems logical, indeed the variance between-organisms decreases as N increases, and also from the mode 1 to the mode 4 of fragmentation, and so selection is less effective at eliminating mutants.

There is no density-dependence in Kondrashov's model: his recurrence equations are written $x_i' = \sum_{j=0}^{N} f_{ij} W_j x_j$ (with constant W_j), and not $x_i' = \sum_{j=0}^{N} f_{ij} \dfrac{W_j}{\overline{W}} x_j$ as in our model. So the matrix $\left[f_{ij}W_j\right]$ is the transition matrix of his system, whereas in our model the transition matrix can be written $f_{ij}W_j/\overline{W}$, $\overline{W}$ changing at each generation. Kondrashov's model is linear, but not ours, and *a priori* we cannot use his technique to calculate the mutation load and the distribution at the equilibrium. Nevertheless, Cushing showed that in this particular case of non-linearity (when the system can be written as the product of a scalar depending on the variables and of a matrix independent of the variables), the equilibrium distribution is still given by the first eigenvector of the matrix [11]. So we obtain the same frequencies at the equilibrium with both models, and thus the same mutation load (this has been checked by simulation).

Otto and Orive add to this model a kind of intraorganismal selection, with $0 < b < 1$ (they don't study the case of selfish mutants with $b > 1$). They show that the mutation load increases with N only if the intraorganismal selection is weak (b close to 1). When the within-organism selection is more important the results are opposite and the load decreases when N increases. One can interpret this by saying that when the inter-organism selection is important compared to the intra-organism selection, the mutants are better eliminated when the variance between-organisms increases, whereas if the intra-organism selection is more important the mutants are better eliminated when the variance decreases (with more mixed groups).

Our preliminary results presented in subsection 3.5 confirm Otto and Orive's results, so long as there is no effect of organism size on fitness. However, when organism size is included in organism fitness mutation load always decreases with decreasing N. It is not clear in the work of Kondrashov [26] and Otto and Orive [42] if the effects on mutation load are indicative of the direction of selection on genes modifying the reproductive system by changing N. However, our results indicate that modifiers for a single-cell stage of the reproductive cycle do the honorable thing and increase when mutation load is decreased.

3.5. Preliminary results. We are currently investigating the equilibrium distribution of propagule states. In Figure 4, the results of simulations of Equation 1 are given assuming binomial sampling as specified in Equation 2 and the mutation selection model specified in Figure 3 and Table 3. The equilibrium frequency distribution appears to qualitatively shift to a U-shaped distribution for small N from a bell-shaped distribution for larger N. The U-shaped distribution is characteristic of loss of intermediate states in which there are mixed groups of the two types of cells, while the bell-shaped distribution is characteristic of maintaining intermediate mixed groups.

This result anticipates a fundamental advantage (at least for selfish mutations) of passing the life cycle through a bottleneck in cell number—increasing the variance at the cell-group level and reducing the within-group variation and conflict. This favors cooperation and helps to restrict the opportunity for defecting mutants to spread. Notice that in Figure 4, the frequency of cooperation is much higher in panel A ($N = 5$) than in panel B ($N = 15$). The limiting case of $N = 1$ corresponds to the transition from fragmentation to spore production. A simple regulatory model has been proposed in which the transition from fragmentation to spore production is caused only by a few changes [48]. The origin of the single cell spore stage is a matter of great importance for individuality and development.

As a first approach to the problem of selection on a modifier reducing propagule size, we consider a population composed of two distinct reproductive types, a "propagule reproducer" and "spore reproducer." Propagule reproducers reproduce according to the model given above (Equation 1 and Equation 2). There are $N+1$ recurrence equations for the propagule reproducer. The spore reproducer reproduces according to the model studied in our previous work and requires just two equations, since there are just two spore types for the haploid case, either C or D. The full vector of state variables is then $[x(0),...,x(N),x(C),x(D)]$, with $x(i)$ = frequency of individuals without modifier with i non-mutant initial cells, $x(C)$ = frequency of C-individuals (with the modifier, starting their development with a non-mutant cell), $x(D)$ = frequency of D-individuals (with the modifier, starting their development with a mutant cell). An individual reproducing via spores starts its development with only one cell and so the number of cells in the adult stage is lower than the number of cells in the adult stage of an individual without the modifier (because t is the same for both kinds of organisms). This can have an effect on fitness, nevertheless we don't include this effect here. We have shown that the frequencies go to an equilibrium given by the first right eigenvector of the $[W_j f_{ij}]$ matrix (constructed in a similar manner to Kondrashov [26] but including the equations for the spore reproducers).

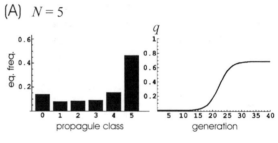

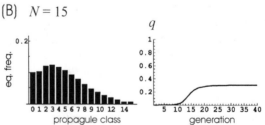

Figure 4. Simulation of propagule model for selfish mutations assuming $W_j = (1 + \beta q_j')k_j/N$. Using Equation 1 and Equation 2. Parameter values in panel (A): $c = 1$, $t = 10$, $\beta = 3$, $\mu = 0.003$, $b = 1.1$; and in panel (B): $c = 1$, $t = 10$, $\beta = 10$, $\mu = 0.01$, $b = 1.05$. As explained elsewhere [40] these parameter values are appropriate for the transition involving cell groups similar in size to some species of Volvocales. The ordinate is either the equilibrium frequency of a propagule class ("eq. freq.") or the frequency of cooperation, q.

When mutations are selfish the spore reproducer always wins, because it is always advantageous to increase the variance between-organisms (and there is no cost of doing so because we have assumed no effect of N on adult size). This is always the case, for both fitness functions in Table 4.

Results for uniformly deleterious mutations are given in Figure 5 for the two fitness functions: when organism fitness is independent of group size (panel (A)) and when organism fitness depends upon group size (panel (B)). When fitness depends only on the frequency of cells (Figure 5 panel (A)), we have found that, when the intraorganismal selection is weak (b close to 1), the spore reproducer is selected, because between-organism selection is more effective at eliminating the deleterious mutations. When intraorganismal selection becomes stronger, the spore reproducer is no longer selected. We have studied the critical value of b for single cell (spore) reproduction to be selected, for several values of the other parameters. In Figure 5, this critical value of b occurs when the $N = 5$ curve intersects the $N = 1$ curve (at a value of b slightly greater than $b = 0.95$ in the figure). The critical value of b doesn't change much when the value of N for the propagule reproducer changes. In other words, the condition for selection of spore reproduction is about the same regardless of the size of the propagules of its competitor. In addition, the mode of fragmentation has a small, but barely noticeable effect, on this critical value. As just mentioned, the curves for other values of $N > 1$ intersect approximately at the same point and behave qualitatively

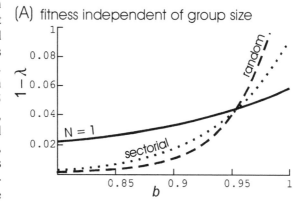

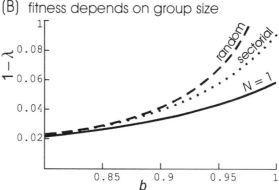

Figure 5. Mutation load ($1 - \lambda$) and selection on spore (stem cell) reproduction for mutations deleterious at both levels of selection. Parameter values are $\mu = 0.003$, $t = 20$, $\beta = 3$. The parameter b on the abscissa is the replication rate of mutant cells. Non-mutant cells replicate at rate 1. Spore production is always selected for selfish mutations ($b > 1$). In panel (A), spore reproduction ($N = 1$) wins when the mutation load decreases with N (to the left of the point where the solid line crosses the others). Key in both panels: dashed lines for $N = 5$ and random mode of fragmentation (Equation 2); dotted lines for $N = 5$, and sectorial mode of fragmentation; solid lines for $N = 1$. The differences between the panels involve the way in which fitness is modeled at the organism (cell-group) level. In panel (A) fitness is independent of group size $W_j = 1 + \beta q_j'$, while in panel (B) fitness depends on group size $W_j = \left(1 + \beta q_j'\right) k_j / N$. See text for explanation.

as does the $N = 5$ curve in that mutation load is greater (or smaller) than the $N = 1$ curve when b is to the right (left) of the intersection. Consequently, the main result concerning the evolution of spore reproduction when fitness is independent of group size is that, when the mutation load increases with N (to the right of the point at which the curves intersect in panel (A) of Figure 5), the modifier is selected (the spore reproducer wins); and when the mutation load decreases with N (to the left of the point of intersection), the modifier is not selected (the propagule reproducer wins).

However, when organism fitness depends upon group size, as it should for simple groups on the threshold of multicellularity, mutation load always decreases with N, and the spore reproducer is always selected (Figure 5 panel (B)). For the values of b lower than 0.8 the value of the load almost doesn't depend on N. Indeed the between-organism selection is reinforced by the effect of size (the more the proportion of deleterious mutants is important, the smaller is the organism). For the values of b lower than 0.8 the proportion of mutants in the population is low, whatever the value of N.

Under the sectorial mode of fragmentation, all cells of a propagule come from the same initial cell in the parent. The kinship among cells in the propagule is higher, and the variance between-organisms is higher, under sectorial sampling than under random sampling. As we see in Figure 5 (panel (A)), when the between-organism selection is more effective at eliminating deleterious mutations (b less than but close to unity), the mutation load is greater with random sampling than with sectorial sampling. The opposite result obtains when within-organism selection is more effective (b significantly less than unity) and the mutation load is lower under random than under sectorial sampling. When fitness depends upon group size (Figure 5 panel (B)), the load is always greater under random sampling than under sectorial sampling.

There are other aspects of the single cell state that require serious consideration, but these have been ignored here. First, there is the opportunity it provides to reset development. The second aspect of the single cell state is its role in the sexual life cycle. It is difficult to imagine sex between-organisms that does not involve a single cell state.[1] We intend to address these viewpoints in due time.

4. Evolution of Conflict Mediation

4.1. Overview. The preliminary results reported in the last section lend credence to the view that, by passing the life cycle through a single cell stage (the stem cell, spore or zygote), the opportunity for between-group selection is enhanced while the opportunity for within-group conflict is reduced. This can favor the transition between single cells and cell groups. We now (and in the next section) consider further aspects of the transition from single cells to organisms by assuming throughout section 4 that cell groups originate from single cells. In our view an organism is more than a group of cooperating cells, more even then a group of cooperating cells related through common

---

[1] Perhaps, bacterial transformation or recombination in viruses (such as occurs during multiplicity reactivation) comes closest to the case in which recombination may occur among DNA molecules from multiple partners. In the slime mold, spores from a single individual may involve mating and recombination of cells from different parents. In the life cycle of the slime mold *Dictyostelium discoideum*, the single cell products of the organism, the spores, form amoebae that ultimately aggregate in groups to form a new multicellular individual, first the migrating slug and then the fruiting body. During the single cell amoebae stage before aggregation, sex (fusion and meiosis) may occur (for review see [50]). As a consequence, a single individual (fruiting body) may involve cells that are recombinations among different parent cells (the degree to which this occurs depends upon the relatedness of the aggregating cells).

descent from a single cell. An organism must have functions that maintain its integrity and individuality. We are interested in understanding the first true emergent level functions during the emergence of the first true individual, the multicellular organism.

All models make assumptions about the processes they study, in the present case, these assumptions involve parameters representing within-organism mutation and cell-cell interaction during development. The purpose of the modifier model introduced here is to study how evolution may modify these assumptions and parameter values, and further how these modifications affect the transition to a higher level of selection. Unlike the classical use of modifier models, say in the evolution of dominance and recombination, the modifiers studied in the present paper are not neutral. Instead, they have direct effects on fitness at the cell and organism level by changing the parameters of within-organism change. By molding the ways in which the levels interact so as to reduce conflict among cells, for example by segregating a germ line early or by policing the selfish tendencies of cells, the modifiers construct the first true emergent organism level functions.

4.2. Model parameters. Recall the parameters of the model studied above. These parameters describe fitness effects at the organism (β) and cell level (b, s_C, s_D), the mutation rate per cell division (μ) and the development time (t). The parameter β is the benefit to organism fitness of cooperation among its component cells. Fitness is assumed to depend linearly (through β) on the number of cooperating cells in the organism's adult stage (see Table 5). The parameter bc is the replication rate of defecting cells (c is the replication rate of cooperating cells, assumed equal to unity). The parameters s_C, s_D are the survivals of C and D cells per cell generation (assumed equal here for simplicity). The parameter μ is the rate of mutation per cell division during development (interpreted here as the genome wide deleterious mutation rate at all loci leading to a loss of cell function). The parameter t is the time available for development (development allows for within-organism change resulting from mutation, μ, and selection, b, s_C, s_D, at the cell level). A new parameter is needed in the case of sexual reproduction with the possibility of recombination between two loci; r is the rate of recombination between the cooperate/defect locus and the modifier locus. Again, the complexity of interaction among different cell types and tissue functions is assumed to be represented by two kinds of interactions—cooperate and not cooperate (or defect). Heritability of fitness at the organism level is measured by the regression of offspring fitness on adult fitness.

A per cell division mutation rate of $\mu = 0.003$ is assumed in many of the studies reported in this section. The reason for assuming this value has been discussed above, in Section 2.3. Nevertheless, smaller mutation rates also result in the evolution of modifiers of within-organism change (that is the transition from equilibrium 3 to 4 discussed below, see, for example, panels (D) and (E) of Figure 6). Higher mutation rates mean more within-organism change and this facilitates evolution of the modifier.

4.3. Two-locus modifier model. We now consider a second locus which is assumed to modify the parameters of within-organism change at the cooperate/defect (C, D) locus (see Table 5 for the additional terms for the modifier model).

The modifier locus is interpreted here as either as a germ line locus or a mutual policing locus, according to whether the modifier allele M affects the way in which cells are chosen for gametes (germ line modifier), or the parameters of selection, and variation at the organism and cell level (policing modifier). A *germ line modifier* is assumed to sequester a group of cells with shorter development time and possibly a lower mutation rate than the soma and less selection at the cell level. A *self-policing modifier* causes the organism to spend time and energy monitoring cell interactions and reducing the advantages of defection at a cost to the organism. In either case, the modifier locus is assumed to have two alleles M and m, with no mutation at the modifier locus. Because there is no mutation at the modifier locus and groups start out as single cells, there can be no within-group change at the modifier locus because there is no within-group variation at the modifier locus. A group of cells expressing allele m is assumed to have the same properties as the cell-groups studied in the previous section (assuming $N = 1$). A group of cells expressing allele M is assumed to have different properties that represent the ideas of a germ line or self policing.

The additional terms and definitions for the two-locus modifier model are given in Table 5. Some explanation of the haploid life cycle considered here may be helpful. In the diploid life cycle, a generation typically begins with the diploid zygote, followed by development, within and between-organism selection, the adult stage, meiosis and the production of haploid gametes which fuse to produce the zygotes of the next generation. In previous analyses of the

Variable	Definition
i, j	indices for genotype 1, 2, 3, 4 = CM, Cm, DM, Dm
k_{ij}	number of i cells in the adult stage (soma) of a j-zygote
k_j	total number of cells in adult stage (soma) of j-zygote, $k_j = \sum_i k_{ij}$
K_{ij}	number of i cells in the germ line of a j-zygote
K_j	total number of cells in the germ line of j-zygote, $K_j = \sum_i K_{ij}$
W_j	individual fitness of j-zygote (assumed to depend on group size): $W_j = k_j + \beta\left(k_{1j} + k_{2j}\right)$
r	recombination rate between C/D and M/m loci
x_j	frequency of two locus j genotype in total population

Table 5. Additional notation and terms for the two locus modifier model. The genotype frequencies, x_j, are measured at the gamete stage, before mating, meiosis, development and within and between-organism selection. The variables, x_j, are now completely different than those defined in the propagule model in Table 4. The table considers germ line modifiers that create a separate germ and somatic line, each of which may have different numbers of cell types (the K_{ij} and k_{ij} variables). In the case of self-policing modifiers, there is no distinction between the germ line and the soma (all cells are potential germ line cells).

two-locus haploid modifier model studied here, it was also assumed that the generation began with the zygote stage (Michod 1996, Michod and Roze 1997). In the haploid life cycle, the haploid zygote stage is followed by development, within and between-organism selection, the adult stage (still haploid), production of haploid gametes, fusion of gametes to produce a transient diploid stage which undergoes meiosis to produce the haploid zygotes of the next generation. Although it doesn't affect the results, the two-locus recurrence equations are simpler for the haploid life cycle, and easier to understand, if we begin a generation with the gametes and let the x_j variables be the frequencies of the different genotypes among gametes (instead of among zygotes).

Consequently, the following sequence of events in each generation is assumed: gametes, fusion of gametes to make the transient diploid stage, meiosis and recombination to form haploid zygotes, development, within and between-organism selection to create the adult stage, and then formation of the gametes to start the next generation.

The recurrence equations for the two-locus haploid life cycle may be constructed as follows. Recombination may change the genotype frequencies according to the level of linkage disequilibrium in the population. *Linkage disequilibrium* measures the statistical association between the frequencies at the two loci. If $G = 0$, the joint distribution of alleles in gametes is the product of the allele frequencies and recombination has no effect on the genotype frequencies. In Equation 3, we define linkage disequilibrium between the locus determining cell behavior and the modifier locus (using the variable G for gametic phase imbalance).

$$G = x_1 x_4 - x_2 x_3 \qquad \text{Equation 3}$$

After meiosis, the frequencies of the four genotypes may change from what they were before fusion, depending on the rate of recombination (r) and the level of linkage disequilibrium, G. The new frequencies after recombination are $x_1 - rG$, $x_2 + rG$, $x_3 + rG$, and $x_4 - rG$, for the CM, Cm, DM and Dm genotypes, respectively. It is upon these new genotype frequencies that selection and mutation operate, as shown in Equation 4. The full two-locus dynamical system is given in Equation 4,

$$x_1' \overline{W} = (x_1 - rG)W_1 \frac{K_{11}}{K_1}$$

$$x_2' \overline{W} = (x_2 + rG)W_2 \frac{K_{22}}{K_2}$$

$$x_3' \overline{W} = (x_3 + rG)W_3 + (x_1 - rG)W_1 \frac{K_{31}}{K_1}$$

$$x_4' \overline{W} = (x_4 - rG)W_4 + (x_2 + rG)W_2 \frac{K_{42}}{K_2}$$

Equation 4

with $\quad \overline{W} = (x_1 - rG)W_1 + (x_2 + rG)W_2 + (x_3 + rG)W_3 + (x_4 - rG)W_4 \quad$ and linkage disequilibrium given by Equation 3. Only three of the equations are independent, since the frequencies must sum to one. Equation 4 gives the same results as the equations studied previously (equations 1 and 2 of Michod 1996 and in Michod and Roze 1997). The primary difference is that genotype frequencies are measured among gametes and not zygotes, and this allows a simpler form of the recurrence equations.

4.4. Equilibria of the system. The transition to a new level of selection proceeds in two general stages. First, cooperation must increase. Deleterious mutation leads to the loss of cooperation. A mutation selection balance may be reached at which the increase of cooperation (by group selection) is offset by its loss through mutation. As discussed previously, the increase of cooperation within the group is accompanied by an increase in the level of within-group change and conflict as mutation and selection among cells leads towards defection and a loss of cooperation [37]. The second stage in an evolutionary transition is the appearance of modifier genes that regulate this within-group conflict. Only after the evolution of modifiers of within-group conflict, do we refer to the group of cooperating cells as an "individual," because only then is the group is truly indivisible in that it possesses higher level functions that protect its integrity.

The main equilibria of the system of are given in Table 6. These equilibria are obtained by setting linkage disequilibrium to zero (G=0, using Equation 3) and by setting the change in genotype frequencies equal to zero (using Equation 4). The equilibria given in Table 6 are the ones we expect based on the biological assumptions of the model, however they assume that linkage disequilibrium (Equation 3) is zero. In the case of germ line modifiers discussed in the next section, we have found examples of other equilibria at which $G \neq 0$, but they are confined to narrow regions of the parameter space. These equilibria with $G \neq 0$ are, nevertheless, quite interesting as the populations are polymorphic at the modifier locus and maintain both germ line and non-germ line phenotypes in the same population.

Eq.	Description of Loci	Interpretation
1	no cooperation; no modifier	Non-functional cell groups, single cells[2]
2	no cooperation; modifier fixed	Not of biological interest, never stable
3	polymorphic for cooperation and defection; no modifier	*Groups of cooperating cells:* no higher level functions
4	polymorphic for cooperation and defection; modifier fixed	*Individual organism*: integrated group of cooperating cells with higher level function mediating within-organism conflict

Table 6. Equilibria for modifier model without linkage disequilibrium. See especially Table A 1 and Table A 2, for a mathematical description of the equilibria and eigenvalues.

For the moment we focus on the equilibria given in Table 6. The eigenvalues of the three independent genotypes $(x_4 = 1 - x_1 - x_2 - x_3)$ are given in Table A 2 for the different equilibria and for sexual and asexual reproduction. The evolution of cooperation corresponds to equilibrium 3 and the evolution of the modifier corresponds to equilibrium 4. Consequently, the question of the transition to individuality boils down to the conditions for a transition from equilibrium 3 to equilibrium 4. The condition for increase of the modifier is given as λ_{31} in Table A 2. In the case of asexual reproduction, $\lambda_{31} > 1$ means that a CM organism must produce more C gametes than a Cm organism.

The evolution of functions to protect the integrity of the organism are not possible, if there is no conflict among the cells in the first place. It is conflict itself (at equilibrium 3) which sets the stage for a transition between equilibrium 3 and 4 and the evolution of individuality.

4.5. Evolution of the germ line. The essential feature of a germ line is that gamete producing cells are sequestered from somatic cells early in development. Consequently, gametes have a different developmental history from cells in the adult form (the soma) in the sense that they are derived from a cell lineage that has divided for a fewer number of cell divisions with, perhaps, a different mutation rate per cell replication and a different selective context. The main parameters influencing the evolution of the germ line is the reduction in development time in the germ line relative to the soma, δ (with the development time in the germ line being $t_M = t - \delta$), the

---

[2] The model assumes cell groups. Nevertheless, we think of Eq. 1 as representing single cells, because there is no cooperation and no interaction between cells to maintain the group structure assumed in the model.

deleterious mutation rate in the germ line, μ_M, and the nature of cell selection in the germ line.

Adult Stage		Genotype of Zygote			
		CM	Cm	DM	Dm
Germ Cells, K_{ij}	CM	$2^{ct_M}(1-\mu_M)^{ct_M}$			
	Cm	0			
	DM	$\dfrac{\mu_M 2^{bct_M} - 2^{ct_M}(1-\mu_M)^{ct_M}\mu_M}{-1+2^{b-1}+\mu_M}$			
	Dm	0			
Somatic Cells, k_{ij}	CM	$2^{ct}(1-\mu)^{ct}$	0	0	0
	Cm	0	$2^{ct}(1-\mu)^{ct}$	0	0
	DM	$\dfrac{\mu 2^{bct} - 2^{ct}(1-\mu)^{ct}\mu}{-1+2^{b-1}+\mu}$	0	2^{bct}	0
	Dm	0	$\dfrac{\mu 2^{bct} - 2^{ct}(1-\mu)^{ct}\mu}{-1+2^{b-1}+\mu}$	0	2^{bct}

Table 7. Numbers of different cell types in the soma and gamete stages for germ line modifiers. Zygote genotype is given across the top (column) and the genotype of the cells after development is given down the rows. For genotypes containing the m allele, there is no germ line; so there is no difference between the germ line stage and the somatic adult stage. The germ line is ignored in D containing zygotes since by assumption there is no mutation from D to C and so no within-organism variation in D containing zygotes. The steps involved in obtaining the formulae in the table are described in Table 3 and Figure 3. We do not consider mutations at the germ line locus, since we are interested in the loss of tissue function and cooperation among somatic cells. A modification of the model given in the text relates the table to the study of a self-policing allele. Selection is allowed in the germ line stemming from the different rates of replication of cooperate and defecting cells. This makes matters more difficult for the origin of the germ line. We also consider in Table 8 the case that, since germ line cells don't cooperate and otherwise function in somatic functions, the replication advantage of defecting cells no longer obtains in the germ line.

We are interested in modeling the origin of germ cells in an ancestor that had none. Furthermore, we wish to consider the hypothesis that the germ line serves to increase the heritability of traits in the organism, as opposed to directly increasing organism fitness. For this reason, we assume that the germ line is costly to fitness (expected number of offspring) although it increases the heritability of fitness (regression of fitness of offspring on fitness of parents). Where do the germ line cells come from? We imagine that the total number of cells is conserved and allocated between the germ line and the soma. Cells sequestered in the germ line are no longer available for somatic function, and, for this reason, the germ line allele may detract from adult organism function. One way of representing this cost is by subtracting the germ line cells from the somatic cells in the adult form (k_{ij} - K_{ij} in Table 7 and Table 8). The number of cells is directly related to the development time, t. So if δ is very small, then the number of cells in the germ line is quite large leaving few cells for somatic function. For this reason, we find that the transition from equilibrium 3 to 4 cannot occur by a continuous increase of δ from zero. There is a threshold value of δ resulting from the cost of the germ line which must be overcome for the modifier to increase. If there were no cost, the germ line allele always spreads in the population.

There is a limitation in our approach to modeling the origin of the germ line. Fitness is assumed to depend on the number of cells in the soma and not on the number of cells in the germ line. Fitness is maximal when no cells are allocated to the germ line, because there are more cells available for somatic function (although the heritability of fitness is less, if there is no germ line). Presumably, the number of cells allocated to the germ line will have some effect on the number of gametes produced, but we have ignored the problem of modeling this effect.

In Table 8 we give the numbers of different cell types in the gamete stages for germ line modifiers, assuming complete differentiation between germ and soma. In other

Adult Stage	Genotype of Zygote CM
CM	$2^{ct_M}(1-\mu_M)^{ct_M}$
Cm	0
DM	$2^{ct_M}\left(1-(1-\mu_M)^{ct_M}\right)$
Dm	0

Germ Cells, K_{ij}

Table 8. Numbers of different cell types in the gamete stages for germ line modifiers assuming no selection in the germ line. See Table 7 for explanation of table. Only the CM zygote is shown here, the other zygotes have germ and soma properties as specified in Table 7.

words, we assume that there is no expression of cooperate/defect phenotype in germ line and so no selection in the germ line ($b = 1$ in the germ line). Deleterious mutation may still occur in the germ line, however it is neutral at the level of the cell in the germ line (mutation still disrupts cell function in the soma, however). We represent no selection in the germ line by assuming $b = 1$ in the table cell corresponding to the DM gametic output of a CM zygote. No selection in the germ line only makes sense for selfish mutations ($b > 1$), since cells that are defective at both levels of selection ($b < 1$) may be expected to replicate more slowly in both the germ and the soma because they are generally defective (not just defective in the cooperative function).

We now return to considering the evolutionary equilibria of the system. The equilibria of Equation 4 correspond to different evolutionary outcomes. Regions of stability of the different equilibria for different parameter values are given in Figure 6 in terms of the reduction of development time caused by the germ line modifier, δ, and the replication rate of mutant cells, b. The replication rate of cooperating cells is unity ($c = 1$). Recall that mutations are beneficial (selfish mutants), neutral or deleterious at the cell level when $b > 1$, $b = 1$, and $b < 1$, respectively.

The transition involving the increase of cooperation (Eq. 1 to Eq. 3 in Figure 6) has been considered previously (Michod 1997a). This transition occurs for parameter values in the regions marked Eq. 3 in the panels in Figure 6, while cooperation will not increase for parameter values in the Eq. 1 region. The transition which interests us here involves the increase of modifiers of within-group change and this transition occurs after cooperation increases (Eq. 3 to Eq. 4 in Figure 6). The conditions under which the population evolves from equilibrium 3 to 4 were studied previously [36]. This transition occurs for parameter values in the Eq. 4 regions in Figure 6. In a later section (Transitions in Individuality), we study the components of this transition in terms of the emergence of fitness and the heritability of fitness at the new organism level.

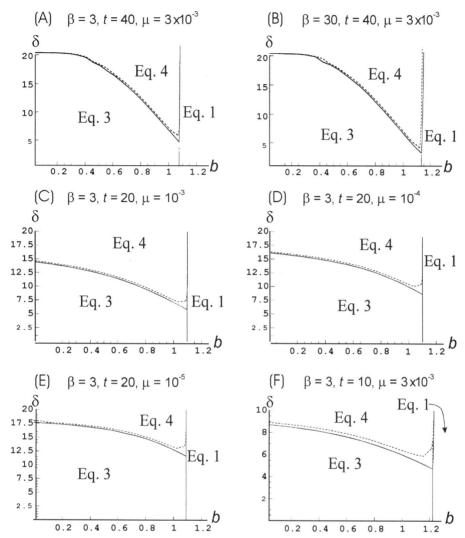

Figure 6. Stability of evolutionary equilibria for germ line modifiers. Results here assume selection in the germ line (Table 7). The modifier is assumed to decrease the development time for the germ line (when compared to the soma) by amount δ. Selfish ($b > 1$) and uniformly deleterious ($b < 1$) mutations are considered. Regions of stability for the different equilibria described in Table 6 are given as a function of the replication rate of mutant cells (relative to cooperating cells), b, and δ, for different values of the mutation rate, μ, development time, t, and advantage of cooperation, β. Solid curves are for asexual reproduction and dashed curves for sexual reproduction assuming a recombination rate of $r = 0.25$ between the modifier and cooperate/defect locus. Cells sequestered in the germ line are not available for somatic function. The mutation rate is assumed to be the same in the soma and the germ line. See also Figure 1 of Michod [37] for more detailed treatment of the boundary between equilibrium 3 and 4 in three dimensions (b, μ, δ).

The general result apparent in Figure 6 is that the modifier increases for all kinds of mutations at the cell level (beneficial, neutral or deleterious) so long as they are deleterious at the organism level (as is assumed throughout) and the reduction in time for replication, δ, is large enough. As the mutants change from being beneficial at the cell level to deleterious, it becomes more difficult for the germ line modifier to increase, in the sense of requiring a larger decrease in the time available for replication in the germ line (all panels in Figure 6). Selfish mutants are most efficacious in selecting for a germ line, nevertheless all mutations deleterious at the organism level promote selection for the germ line modifier.

The parameters have understandable effects on the regions of stability of the different equilibria described in Table 6. For example, as the benefit of cooperation to the group increases from $\beta = 3$ to $\beta = 30$, larger replication benefits of defection at the cell level are tolerated as shown in panels (A) and (B) of Figure 6. Likewise, as the size of the organism decreases from $t = 40$ to $t = 20$ to $t = 10$ (shorter development times), larger benefits of defection at the cell level are tolerated although the reduction in development time for the germ line to evolve is about the same (see panels (A), (C) and (F) of Figure 6). As the mutation rate decreases from 10^{-3} to 10^{-4} to 10^{-5}, it becomes more difficult for germ line modifiers to increase, in the sense that larger reductions in development time are required for there to be a transition from equilibrium 3 to 4 (see panels (C), (D) and (E) of Figure 6; see also Figure 1 of Michod [37]). Higher mutation rates mean more within-organism change and this makes it easier for conflict modifiers to increase.

Figure 6 shows the limits of these regions of stability of the four equilibria in Table 6 in (b, δ) space, for different values of the time for development, t, the benefit of cooperation, β, and μ the mutation rate. Recall, the parameter δ is the difference between the development time of the soma (t) and of the germ line. In the case of asexual reproduction these different regions don't overlap. With sexual reproduction, there are regions in which more than one equilibrium can be stable at the same time (bi-stability) and regions in which none of the four equilibria given in Table 6 are stable. Such a situation is shown in Figure 7, for selfish mutations ($b > 1$).

In Figure 7, there is assumed to be no selection in the germ line (Table 8), although the conclusions discussed now also apply when there is selection in the germ line (Table 7). There are three curves given in Figure 7 which overlap in different regions. First there is a nearly vertical dotted line at about $b = 1.1$. To the right of this line, equilibrium 1 is stable and there is no cooperation among cells. Second there is a solid curve which defines the region of stability for equilibrium 3. Below this curve equilibrium 3 is stable and there is cooperation but the germ line modifier won't increase. Third, there is a dashed-dotted curve which defines the region of

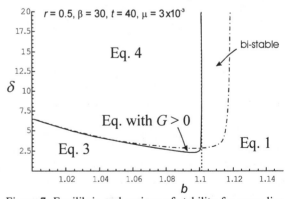

Figure 7. Equilibria and regions of stability for germ line modifier and selfish mutations. No selection in germ line (Table 8), although the qualitative relationships of the three curves (bi-stable and Eq. with $G > 0$ regions) are similar if there is selection in the germ line (Table 7), or if the recombination rate is smaller so long as $r > 0$. See text for explanation.

stability for equilibrium 4. Above this curve equilibrium 4 is stable; both cooperation and the germ line modifier are present.

The region in which both equilibrium 1 and 4 are stable is labeled bi-stable in Figure 7. Depending on initial conditions the population can switch between no cooperation to a fully functional organism. This requires the introduction of the two alleles C and M at the same time and on the same chromosome. We consider this to be a rare and unlikely event. Nevertheless this suggests that with sex, mixed populations of individual cells (no organisms, Eq. 1) and organisms, interpreted as well integrated groups of cells (Eq. 4), may coexist. However, the transition from equilibrium 1 to equilibrium 4 is less interesting in terms of an evolutionary scenario towards individuality, because it supposes the simultaneous appearance of C and M alleles in the population (cooperation and germ line). It is more reasonable to consider the evolutionary transition via equilibrium 3. Consequently, the boundary which interests us is the one between the region of stability of equilibrium 3 and of equilibrium 4.

In Figure 7, the curves defining the regions of stability of equilibrium 3 and 4 are disjunct for a narrow region of parameter space (approximately δ around 2.6 and b in between 1.06 and 1.1). This region is labeled "Eq. with $G > 0$." In this region none of the four equilibria with zero linkage disequilibrium are stable. There is, however, in this region a stable equilibrium with positive linkage disequilibrium, so that the CM and Dm chromosomes are more common than predicted by the product of the allele frequencies. For parameters in this region, selection and recombination are balanced at the $G \neq 0$ equilibrium. Selection is favoring the CM chromosome and disfavoring the Dm chromosome. However, recombination is breaking these chromosomes apart generating Cm and DM chromosomes, configurations for which the modifier is either not present (Cm) or has no effect (DM).

We have only recently discovered the existence of these equilibria with positive linkage disequilibrium. Although the range of parameters permitting this equilibrium are so far small, the $G > 0$ equilibrium interests us greatly since the reproductive system is polymorphic in the population; both organisms with and without germ lines coexist in such equilibrium populations. An example of this is given in the computer simulation in Figure 8. As just mentioned there is a stable polymorphic equilibrium at the germ line locus so that both germ line and no germ line phenotypes are maintained in the population.

In summary, the transition to individuality via equilibrium 3 involves two steps: initial increase of cooperation within the group, and concomitantly of the level of within-group change since mutation leads to loss of cooperation, and then appearance of the germ line to regulate this within-group conflict. Only after the evolution of modifiers of within-group conflict, can we refer to the integrated group of

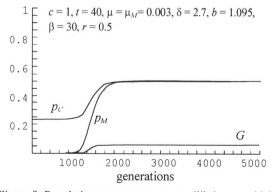

Figure 8. Population converges to an equilibrium at which $G > 0$ and is polymorphic for the germ line modifier. The variables p_C, p_M and G are the frequencies of cooperation, the modifier allele, and linkage disequilibrium, respectively. See Figure 7 and text for explanation.

cooperating cells as an "individual," since the group is now truly indivisible and possesses higher level functions that protect its integrity.

4.6. Evolution of the mutation rate. Modifiers lowering the mutation rate ($\mu_M <$ μ) are also selected for in this model. Maynard Smith and Szathmáry suggest that germ line cells may enjoy a lower mutation rate but do not offer a reason why [32]. Bell interpreted the evolution of germ cells in the Volvacales as an outcome of specialization in metabolism and gamete production to maintain high intrinsic rates of increase while algae colonies got larger in size ([4], see also Maynard Smith and Szathmáry [32] pp. 211-213). We think there may be a connection between these two views.

As metabolic rates increase so do levels of DNA damage. Metabolism produces oxidative products that damage DNA and lead to mutation. It is well known that the highly reactive oxidative by-products of metabolism (for example, the superoxide radical O_2^-, and the hydroxyl radical $\cdot OH$ produced from hydrogen peroxide H_2O_2) damage DNA by chemically modifying the nucleotide bases or by inserting physical cross-links between the two strands of a double helix, or by breaking both strands of the DNA duplex altogether. The deleterious effects of DNA damage make it advantageous to protect a group of cells from the effects of metabolism, thereby lowering the mutation rate within the protected cell lineage.

This protected cell lineage—the germ line—may then specialize in passing on the organism's genes to the next generation in a relatively error free state. Other features of life can be understood as adaptations to protect DNA from the deleterious effects of metabolism and genetic error [35]: keeping DNA in the nucleus protects the DNA from the energy intensive interactions in the cytoplasm, nurse cells provision the egg so as to protect the DNA in the egg, sex serves to effectively repair genetic damage while masking the deleterious effects of mutation. The germ line may serve a similar function of avoiding damage and mutation—by sequestering the next generation's genes in a specialized cell lineage these genes are protected from the damaging effects of metabolism in the soma.

As just mentioned, according to Bell [4], the differentiation between the germ and the soma in the Volvocales results from increasing colony size, with true germ soma differentiation occurring only when colonies reach about 10^3 cells as in the *Volvox* section *Merillosphaera*. Assuming no cell death, a colony size of 10^3 cells would require a development time of approximately $t = 10$ in our model (see panel (F) of Figure 6 for the case $t = 10$; in reality, because of cell death, larger t with more risks of within colony variation would be needed to achieve the same colony size). Although Bell interpreted the dependence of the evolution of the germ line on colony size as an outcome of reproductive specialization driven by resource and energy considerations, this relation is also explained by the need for regulation of within colony change. Colony size increases as t increases, but so does the opportunity for conflict and the need to regulate within-group change.

4.7. Evolution of self-policing. We now consider another means of reducing conflict among cells, that of self-policing. In the model analyzed now there is assumed to be no germ line, although presumably a germ line and self-policing may operate together, that case is not explicitly studied here. Organisms may reduce conflict by actively policing and regulating the benefits of defection [7,17]. How might organisms police the selfish tendencies of cells? The immune system and programmed cell death are two possible examples. There are several introductions to the large and rapidly developing area of programmed cell death, or apoptosis [1,2,10]. To model self-policing, we let the modifier allele affect the parameters describing within and between-organism selection and the interaction among cells. Within-organism selection is still assumed to result from differences in replication rate, not cell survival, by assuming $s_C = s_D = 1$. Cooperating cells in policing organisms spend time and energy monitoring cells and reducing the advantages of defection to $b - \varepsilon$ at a cost to the organism, δ. The parameter δ is now completely different from the germ line modifier δ; δ is now the fitness cost of self-policing at the organism level. To sum up, m genotypes are described by the parameters b and β, while in M genotypes, the benefit of cooperation becomes $\beta - \delta$ and the benefit of defection becomes $b - \varepsilon$.

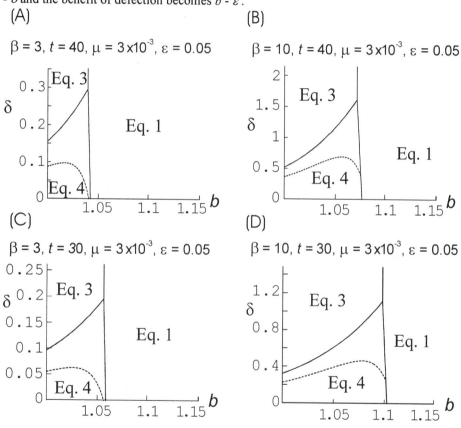

Figure 9. Stability of evolutionary equilibria for self-policing modifiers. Regions of stability of the different equilibria as a function of the advantage of defection, b, and the cost of policing, δ, for different values of the development time, t, and benefit of cooperation to the organism, β. The equilibria are described in Table 6. The modifier is assumed to decrease the advantage of defection by amount ε to $b - \varepsilon$ ($\varepsilon = 0.05$ in all panels). Solid curves are for asexual reproduction and dashed curves for sexual reproduction assuming a recombination rate of $r = 0.25$. See Figure 3 of Michod [36] for more detailed treatment of the boundary between equilibrium 3 and 4 in three dimensions (b, t, δ).

Figure 9 shows the regions of stability of the different equilibria given in Table 6 as a function of b and δ, for several values of development time, t, and benefit of cooperation at the whole organism level, β. The modifier increases (Equilibrium 4 is stable), if the cost of policing, δ, doesn't exceed the boundary between regions Eq. 3 and Eq. 4 in Figure 9.

In Figure 9, there is a threshold level of the benefit of defection, b, above which the organism cannot be maintained, with or without the modifier; equilibrium 1 is stable in this region. This region is defined as the nearly vertical line defining Eq. 1 in Figure 9. This threshold increases, permitting greater levels of defection, as the development time decreases (compare panel (C) with panel (A) and panel (D) with panel (B)). Once the modifier evolves (region Eq. 4), greater levels of defection are tolerated as the threshold slants to the right. The effect is more pronounced for higher levels of β ($\beta >$ 10; results not shown, but one can see the general effect by comparing panels (B) and (A) of Figure 9). This effect also occurs in the case of the germ line modifier (see panel (B) of Figure 6).As the benefit of defection begins increasing, larger costs of policing are tolerated and the modifier still increases (boundary between Eq. 3 and Eq.4 tends upward as b increases from 1). Recombination (dashed curve) reduces the prospects for the policing modifier, as it did in the case of germ line modifiers (Figure 6), although the effect is larger in magnitude in the case of policing modifier. This effect of recombination becomes most pronounced as the Eq. 1 threshold is reached (Figure 6 and Figure 9), leading to the humped curve defining the boundary between Eq. 3 and Eq. 4.

There are important differences in how self-policing and germ line modifiers are modeled. In the case of the germ line, both the cost and the benefit of the M allele vary with δ, the reduction in development time in the germ line. The cost of the germ line increases as δ decreases. In the case of self-policing, the cost and the benefit are independent. In the graphics of Figure 9, the benefit of defecting at the cell level is fixed at $\varepsilon = 0.05$ (the replication rate of the D cells is lowered by 5%), while the cost to the organism of the policing defecting cells, δ, is given on the y-axis.

4.8. Summary of modifier evolution. Modifiers increase by virtue of being associated with more fit genotypes and by increasing the heritability of fitness of these types. For example, at Eq. 3, cooperating zygotes are more fit than defecting zygotes. The cooperating groups must be more fit, because for equilibrium 3 to be stable the fitness of groups with cooperators must compensate for directional mutation towards defection (from C to D). This is what the eigenvalue conditions tell us, as can be seen by considering the eigenvalues in Table A 2. As discussed further elsewhere [39], the eigenvalues in Table A 2 are products of cell-group (organism) fitness and heritability at the cell group level (heritability decreasing with the amount of within-group change). It can be seen (Table A 2) that we need $W_2 > W_4$, for Eq. 3 to be stable ($\lambda_{33} < 1$) and Eq. 1 unstable ($\lambda_{12} = 1/\lambda_{33} > 1$) in the single locus case assuming asexual reproduction. Modifiers increase by virtue of increasing the heritability of fitness of the already more fit type and by hitchhiking along with these more fit chromosomes. They increase the heritability of fitness of the already more fit type by decreasing the within-group change.

For the modifiers to increase, it is not necessary for cooperation per se to exist, only that there be a fitness difference between genotypes and further that the modifiers increase the heritability of fitness of the favored type. For example, if we considered a parallel situation of mutation from defection to cooperation (from D to C), the defection equilibrium (Eq. 1) would involve a mixture of types (just like Eq. 3). For equilibrium 1 to be stable in the first place (before the modifier is introduced) D groups must be more fit than groups composed of cooperators (more fit because of the assumed size advantage of D groups). We suspect that germ line modifiers would then increase, again by increasing the heritability of fitness, but this time heritability of fitness of D zygotes. However, in this situation, even though the modifiers increase, we would not want to speak of there being a transition to a new higher level individual or unit of selection. There cannot be a new individual, or unit of selection, at the level of the group without there being interactions among its members by virtue of which the group gains new functionality.

5. Transitions in Individuality

5.1. Effect of transition on the level of cooperation. We now consider the consequences of an evolutionary transition from equilibrium 3 to 4 on the level of cooperation and synergism attained (this subsection) and on the heritability of fitness at the new organism level (in the next two subsections). For reasons of space, we only consider the evolution of the germ line, but qualitatively similar results have been obtained for the other forms of conflict mediation such as self-policing and determinate size. We also assume the case of cell selection in the germ line (Table 7), although similar results exist if there is no selection in the germ line (Table 8).

In Figure 10 the frequencies of the

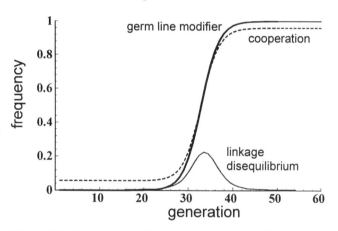

Figure 10. Frequencies of cooperation and modifier during evolutionary transition. Germ line modifier refers to the M allele and cooperation to the C allele; asexual reproduction is assumed. The values of the parameters are $c=1$, $t=40$, $\delta=35$, $b=1.1$, $\beta=30$, $r=0$, $\mu=\mu_M=0.003$. These parameter values were chosen to illustrate the components of an evolutionary transition, as they produce large changes in the frequencies before and after the transition. However, as shown in Figure 6 (and Figure 9 for self-policing modifiers) a transition occurs for all parameter combinations in the "Eq. 4" regions of the different panels.

cooperation and modifier allele are plotted along with the linkage disequilibrium. We see that the transition dramatically increases the level of cooperation in the population, and that during the transition the coupling chromosomes (CM and Dm) predominate. The level of cooperation always increases during a transition, whatever the values of the parameters (if equilibrium 4 is stable). To understand the effect of this evolutionary transition on the regulation of the within-organism change and the heritability of fitness at the new level, we need to adapt the covariance methods of Price to the present system of equations.

5.2. Increase of fitness covariance at organism level.

5.2. Increase of fitness covariance at organism level. The recurrence equations above are derived by directly monitoring the numbers and frequencies of cells at the different life stages. An alternative method for representing selection in hierarchically structured populations is Price's covariance approach [43-45]. The covariance approach to the present situation is discussed in more detail elsewhere [37,39,40]. Price's approach posits a hierarchical structure in which there are two selection levels—in our case, (i) between cells within-organisms—viewed as a group of cells—and (ii) between-organisms within populations. Gene frequency change at the cooperate/defect locus is given in Equation 5,

$$\Delta q = \frac{\mathrm{Cov}_{\mathbf{q}}[W_i, q_i]}{\overline{W}} + \mathrm{E}_{\mathbf{Wq}}[\Delta q_i], \qquad \text{Equation 5}$$

with the following vectors used as weights $\mathbf{q} = (1-q, q)$, $\mathbf{Wq} = (W_D(1-q), W_C q)$. Variables q and q_i are the frequencies of a gene of interest in the total population and within zygotes; $\mathrm{Cov}_{\mathbf{q}}[x, y]$ and $\mathrm{E}_{\mathbf{Wq}}[x]$ indicate the weighted covariance and expected value functions respectively. The Price covariance Equation 5 partitions change to the two levels of selection. The first term of the Price equation is the covariance between fitness and genotype and represents the heritable aspects of fitness; the second term is the average of the within-organism change. The first term can be considered as representing the selection between-organisms within the population, and the second term the selection between cells within the organism. When the population is at an equilibrium, $\mathrm{D}q = 0$ and so it must be the case that the two terms on the right hand side of Equation 5 equal one another in magnitude $\dfrac{\mathrm{Cov}_{\mathbf{q}}[W_i, q_i]}{\overline{W}} = -\mathrm{E}_{\mathbf{Wq}}[\Delta q_i]$.

In Figure 11, the two components of the Price covariance Equation 5 are plotted during the transition from equilibrium 3 to equilibrium 4 given in Figure 10 during the increase in frequency of the germ line allele, from 0 to 1 [43]. These components partition the total change in gene frequency into heritable fitness effects at the organism level (solid line) and within-organism change (dashed line). In the model studied here, within-organism change is always negative, since defecting cells replicate faster than cooperating cells and there is no back mutation from defection to cooperation. At equilibrium, before and after the transition, the two components of the Price equation must equal one another. This can be seen in Figure 11 by the fact that the two curves begin and end at the same point. However, during the transition we see that the covariance of individual fitness at the emerging organism level with zygote genotype (solid curve of Figure 11) is greater than the average change at the cell level (dashed curve of Figure 11).

This greater covariance in fitness at the higher level forces the modifier into the population. In Figure 11, we see that modifiers of within-organism change evolve by making the covariance between fitness at the organism level and zygote genotype more important than the average within-organism change. This implies that modifiers increase the heritability of fitness at the new level.

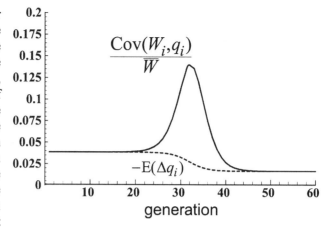

Figure 11. Study of evolutionary transition by Price equation. Same parameter values as Figure 10. Figure adapted from Michod and Roze [40].

5.3. Heritability of fitness and the evolution of individuality. Darwin argued that natural selection requires heritable variations in fitness [12]. Levels in the biological hierarchy—genes, chromosomes, cells, organisms, kin groups, groups—posses heritability of fitness to varying degrees according to which they may function as evolutionary individuals, or units of selection [24,28]. Beginning with Wilson [54] and the study of the transition from solitary animals to societies, then Buss [9] with the study of the transition from unicellular to multicellular organisms, and more recently Maynard Smith and Szathmáry [32,49], attention has focused on understanding transitions between these different levels of selection or different kinds of evolutionary individuals.

Before the evolution of cooperation, in the present model, the population is composed of Dm cell types (equilibrium 1 in Table 6). In such a population the heritability of fitness equals unity, because there is either no sex, or no effect of recombination if there were sex (in $Dm \times Dm$ matings), and there is no within-organism variation or change (we assume no mutation from D to C). When the C allele appears in the population, evolution (directed primarily by kin selection) may increase its frequency leading to greater levels of cooperation (from equilibrium 1 to equilibrium 3). With the evolution of greater cooperation, within-organism change increases, because of mutation from C to D and selection at the cell level. As a consequence of the evolution of cooperation, and increasing within-organism change, the heritability of fitness must decrease.

The organism cannot evolve new adaptations, such as enhanced cooperation, if these adaptations are costly to cells, without increasing the opportunity for conflict within and thereby decreasing the heritability of fitness. Deleterious mutation is always a threat to new adaptations by producing cells that go their own way. By regulating within-organism change, there is less penalty for cells to help the organism. Without a means to regulate within-organism change, the "organism" is merely a group of cooperating cells related by common descent. Such groups are not individuals, because they have no functions that exist at the new organism or group level.

The existence of a zygote stage in the life cycle serves to decrease the within-organism change by increasing the relatedness among cells. However, as showed elsewhere [37], within-organism change can be significant even in this case. The main criteria of significance is whether within-organism variation leads to selection of modifiers to reduce it. We have found that such modifiers increase in frequency leading to an evolutionary transition that we have interpreted as individuality because these modifiers represent the first higher level functions. However, does heritability of fitness—the defining characteristic of an evolutionary individual—actually increase during the transition between equilibrium 3 and 4?

Heritability of fitness at the new cell group or organism level may be defined as the regression of the fitness of offspring cell groups on fitness of the parent cell groups (see, for example, reference [16]). It can be shown that when the population is at equilibrium 3 or at equilibrium 4, this definition gives a simple expression for heritability equal to $h_W^2 = \dfrac{k_{22}}{k_2}$ at equilibrium 3 and equal to $\dfrac{K_{11}}{K_1}$ at equilibrium 4. We always have $\dfrac{K_{11}}{K_1} > \dfrac{k_{22}}{k_2}$, so the evolutionary transition always leads to an increase in the heritability of fitness. If we go back to the eigenvalues of the different equilibria given in Table A 2 we can see that these eigenvalues are ratios between products of fitnesses and heritabilities. This illustrates clearly that what determines if a new characteristic can increase in frequency in the population is the heritability of fitness of individuals with the new feature.

During the transition between equilibrium 3 and equilibrium 4, all four genotypes are present in the population. It's not possible to simplify the heritability as above, however the following expression for heritability may still be used $h_W^2 = \dfrac{\text{Cov}(W_P, W_O)}{\text{Var}(W_P)}$, where W_P is the fitness of each parent cell group and W_O is average fitness of the offspring cell groups produced by the parents.

Initially, before the evolution of cooperation between cells, the heritability of fitness is unity. After cooperation evolves, because of high kinship, heritability is significant at the group (organism) level ($h_W^2 \approx 0.6$, Figure 12), but this value is still low for asexual haploidy (heritability at the organism

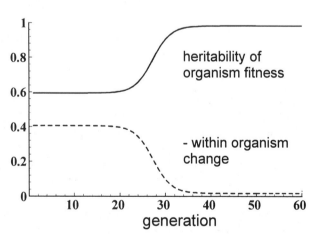

Figure 12. Heritability of organism fitness and within-organism change during evolutionary transition. Same parameter values as Figure 10. Figure adapted from Michod and Roze [40].

level should equal unity in the case of asexual organisms when there is no environmental variance). Low heritability of fitness at the new level resulting from significant within-organism change posses a threat to continued evolution of the organism. In the case considered in Figure 12, development time, and hence organism size, could not increase without the evolution of conflict modifiers. Indeed, as already noted, the continued existence of cell-groups at all is highly unlikely, since the cooperation allele is at such a low frequency and stochastic events would likely lead to its extinction. Before the evolution of modifiers restricting within-organism change, the "organism" is just a group of cooperating cells related by common descent from the zygote. As the modifier begins increasing, the level of within-organism change drops (dashed curve Figure 12) and the level of cooperation among cells increases dramatically (dashed curve of Figure 10) as does the heritability of organism fitness (solid curve of Figure 12).

The essential conclusion is that even in the presence of high kinship among cells, there remains significant within-organism change, by "significant" we mean this change leads to the evolution of a means to regulate it, such as the segregation of a germ line during the development or the evolution of self-policing. Once within-organism change is controlled, high heritability of fitness at the new organism level is protected. Individuality at the organism level depends on the emergence of functions allowing for the regulation of conflict among cells. Once this regulation is acquired, the organism can continue to evolve new adaptations without increasing the conflict among cells, as happened when cooperation initially evolved (transition from equilibrium 1 to equilibrium 3).

6. Conclusions

The models studied here support the view that single cell or spore reproduction, the germ line, self policing and determinant size evolved to increase the heritability of fitness and to help mediate conflict between cooperating and defecting cells. As a consequence, these adaptations served to facilitate a transition between cells and multicellular organisms. Development evolves, at least during its initial phases, so as to reduce the opportunity for conflict among cells. Having a germ line functions to reduce the opportunity for conflict among cells and promote their mutual cooperation both by limiting the opportunity for cell replication [9] and by lowering the mutation rate [32]. Mutual policing [7,17] is also expected to evolve as a means of maintaining the integrity of the organisms once they reach a critical size. Any factors that directly reduce the within-organism mutation rate are also favored. These modifiers of within-organism change during development increase by virtue of being associated with the cooperating genotype, which is more fit than the defecting genotype at the time when the modifiers are introduced (at equilibrium 3). The modifiers increase the heritability of fitness of the more fit type, in our case the cooperating type.

We have only recently discovered cases in which the population remains polymorphic for two reproductive strategies, one involving conflict mediation and the other not (for example, the polymorphic germ line equilibria in Figure 7 and Figure 8). These cases are of interest to us, because of their possible relevance to the mixed reproductive mode so common in plants, in which vegetative (propagule) reproduction and seed (or spore) reproduction are maintained together in populations. As studied in section 3, seed (or spore) reproduction can be seen as a mechanism to reduce conflict and can evolve in populations reproducing by fragmentation. Furthermore, the mode of reproduction has profound implications for mutation load and the distribution of mutations. The growth habits of plants are more indeterminate than animals and plant fitness probably often depends upon organism size and the replication rates of component cells. Furthermore, levels and mechanisms of within-organism change are well documented in plants [3,14,18,25,53], and the models studied here seem especially relevant. It is often noted that plants don't have a germ line. The consequences of within-organism change are not as severe in plants as in animals, because plant cells cannot move. Because of the presence of a rigid cell wall in plants, the opportunity for systematic infections of cancerous cells is severely reduced [9].

Godelle and Reboud consider a class of two level selection models that, although different from ours in orientation and purpose (they consider primarily segregation distortion and do not explicitly interpret their results in terms of evolutionary transitions), have some similar properties and comparable results [20,21]. A single diploid locus is considered affecting between-organism and within-organism selection. Within-organism change is represented by segregation distortion in the heterozygote. A gene frequency equation is derived giving the total gene frequency change through a generation accounting for selection at both levels and, in addition, for inbreeding. Through analysis of the gene frequency equation and computer tournaments between invading mutants and resident alleles the evolutionary dynamics are characterized in terms of replacement of alleles by new mutations. In general [21] no constraints on the effects of the mutants on the two levels of selection are considered, although in [20] there is considered to be a trade-off between the performance of a mutation at the two levels. In reference [21] a "synthetic fitness function" is proposed that is maximized during the course of evolution and is a product of the fitnesses at the two levels. Inbreeding serves to favor between-organism selection and reduce within-organism change.

In our model modifiers of within- and between-group change are introduced at an equilibrium that is polymorphic for two kinds of group, those stemming from cooperating or defecting zygotes. By definition, cooperative genotypes bias the selection towards the group (organism) level (by taking fitness away from cells while increasing the fitness of the group), while defecting genotypes do the opposite and bias selection toward the cell level. To maintain cooperation (at equilibrium 3) cooperative groups must be more fit (produce more gametes) than defecting groups, because the fitness benefits of cooperation at the group level must compensate for mutation towards defection.

As discussed in subsection 4.8, modifiers of within and between-organism change increase by virtue of being associated with the more fit type (cooperation) when they are introduced and by increasing the heritability of fitness of this type. As a result, the modifiers increase the heritable component of fitness at the organism level—that is the product of heritability and fitness. This product function is similar to the synthetic fitness function of Godelle and Reboud in which the product of selection at the two levels is taken [21]. In reference [21], selection within the organism is due to segregation distortion, which can be viewed as reducing the heritability of the genotypic properties of the heterozygote (the heterozygote begins with an equal ratio of the two alleles but produces a different ratio among its gametes).

Heritability is also a simple function of the level of within-organism change in our model. The question of a transition to individuality reduces in our model to the conditions for a transition from equilibrium 3 to equilibrium 4 in Table 6 (cooperation with and without the modifier). The condition for increase of the modifier in the case of asexual reproduction is given as λ_{31} in Table A 2. The condition $\lambda_{31} > 1$ simply means that a cooperating organism must produce more C gametes with the modifier (CM) than without (Cm). As discussed elsewhere [39] the eigenvalue λ_{31} is a comparison of product of fitness and heritability for the CM and Cm genotypes $\left(W_1 K_{11}/K_1\right)/\left(W_2 k_{22}/k_2\right)$. In our model, heritability is inversely related to the level of within-organism change and equals K_{11}/K_1 and k_{22}/k_2 after and before the transition, respectively (see Figure 6-5 of [39]). As already mentioned in the last subsection, we always have $K_{11}/K_1 > k_{22}/k_2$, so the evolutionary transition leads to an increase in the heritability of fitness.

An evolutionary transition is diagrammed in Figure 13 for selfish

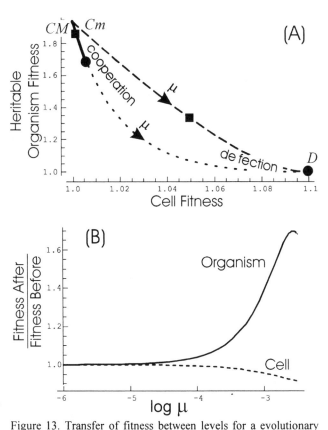

Figure 13. Transfer of fitness between levels for a evolutionary transition involving selfish mutations ($b = 1.1$). Same parameter values as in Figure 10, Figure 11, and Figure 12 except that in both panels the mutation rate μ is changing. Panel (A) is a parametric plot while panel (B) is a standard plot. In panel (A) the dashed line is for equilibrium 3 before the transition (Cm) and the solid-dotted line is for equilibrium 4 after the transition (CM). The modifier has no effect in defecting cells. Organism fitness depends on group size and functionality (Table 5). Average heritable organism fitness (expressed relative to defecting groups) is shown on the ordinate in panel (A) and is calculated as $\left(W_1 f_{11} x_1 + W_3(1 - x_1)\right)/W_3$ after the transition and $\left(W_2 f_{22} x_2 + W_4(1 - x_2)\right)/W_4$ before. Because the heritable fitness of cell groups is expressed relative to defecting groups, the heritable fitness of defecting groups is unity at all points in the figure. Average cell fitness is shown on the abscissa in panel (A) and is calculated as $cf_{11} x_1 + b(1 - x_1) + b(1 - f_{11}) x_1$ after the transition and $cf_{22} x_2 + b(1 - x_2) + b(1 - f_{22}) x_2$ before. Ratios of these fitnesses at the two levels are plotted on the ordinate in panel (B). See text for explanation. In panel (A), the points of equal mutation rates are indicated by solid squares for $\mu = 10^{-3}$, and solid circles for $\mu = 3.16 \times 10^{-3}$.

mutations with regard to the spectrum of fitness variation at the organism and cell levels before and after the transition as a function of the mutation rate, μ. The same parameter values are assumed in Figure 13 as in Figure 10, Figure 11, and Figure 12, except that in Figure 13 in both panels the mutation rate is not fixed but μ ranges from 10^{-6} to 3.16×10^{-3} (at which point equilibrium 3 no longer exists). In panel (A) the average heritable component of fitness (heritability times expected number of gametes) of the cell group, the organism, is plotted on the ordinate and the average fitness of the cell is plotted on the abscissa as a function of the mutation rate , μ. The calculation of these average heritable components of fitness is explained in the legend to the figure. Two curves are plotted parametrically in panel (A) as a function of the mutation rate. The top curve corresponds to the situation before the modifier increases (dashed line) and the bottom curve to the situation after the transition occurs (solid - dotted line). In both cases, as the mutation rate increases from near zero ($\mu = 10^{-6}$) to a high level, the population shifts from being primarily composed of cooperating groups (*CM* zygotes for the bottom curve and *Cm* zygotes for the top curve) to defecting groups (*D* zygotes). In other words, as the mutation rate increases, the fitness at the organism level must decrease and the fitness at the cell level must increase. This, after all, is the definition of selfish mutations (considered in the figure). However, the rate and manner in which this transfer of fitness occurs is quite different before and after the transition.

Before the transition (top dashed curve) the population shifts from nearly complete cooperation at $\mu = 10^{-6}$ to complete defection $\mu = 3.16 \times 10^{-3}$ (at which point equilibrium 3 no longer exists). For this same increase in the rate of mutation, the population after the transition (bottom curve in panel (A)) changes only a small amount in terms of its fitness at the two levels, as is shown by the solid portion of the curve (the solid circles correspond to the fitness at the two levels for a mutation rate of $\mu = 3.16 \times 10^{-3}$). Only the solid portion of the bottom curve is comparable to the entire dashed curve before the transition (panel (A) of Figure 13). In other words with conflict mediation in place the deleterious effects on fitness of selfish mutations are buffered for the organism. The rest of the bottom curve (the dotted portion) is generated by letting the mutation rate increase past 3.16×10^{-3} into ranges not permitted before the transition. The two curves must begin and end at approximately the same point because the modifier has little effect either when there is no mutation or a lot of mutation.

A study similar to that in Figure 13 where $b = 1.1$, can be made for uniformly deleterious mutations and this is done in Figure 14 for $b = 0.9$. As shown in Figure 6, the modifier increases in such cases (panel (B) of Figure 6 is closest to the situation studied in Figure 13 and Figure 14). The curves in panel (A) have a positive slope in Figure 14, because there is no longer any conflict between the fitness effects of mutation at the two levels. Mutation is deleterious at both levels. There is a dramatic effect of mutation on the fitness of the cell group, because mutation not only reduces the functionality of the group but decreases its size as well.

The ordinate axis in panels (A) of Figure 13 and Figure 14 are vastly different. Why does the heritability of fitness at the organism level change more drastically with the mutation rate, μ, when mutations are uniformly deleterious (as in Figure 14) than when they are selfish (as in Figure 13)? A given decrease in average cell fitness (the abscissa of panel A in both figures) has conflicting effects at the organism level in Figure 13 but not in Figure 14. Reducing the average deleterious nature of the mutations (moving from left to right along the abscissa of panel (A) of Figure 14), dramatically increases the fitness of the organism by increasing its size as well as its functionality. On the other hand, reducing the advantage of defection (moving from right to left along the abscissa of panel (A) of Figure 13) has conflicting effects on organism fitness because it increases the functionality of the organism but also decreases its size. To be explicit, consider, the case with near zero mutation with little or no within-organism change. For selfish mutations the ratio of fitness of cooperating and mutant groups is about 1.9 (the upper left-hand corner of panel (A) Figure 13), while for uniformly deleterious mutations ($b = 0.9$) the ratio of fitness of cooperating and mutant groups is about 500 (the upper right-hand corner of panel (A) in Figure 14). Again, this is because with uniformly deleterious mutations the mutant groups are much smaller in size as well as being less functional than cooperating groups, while with selfish mutations the mutant groups are larger in size than cooperating groups (but less functional).

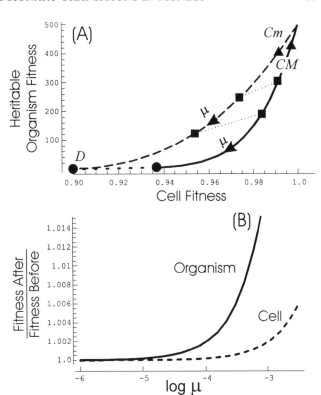

Figure 14. Fitness effects of increase of modifiers for uniformly deleterious mutations ($b = 0.9$). Same legend as Figure 13, except $b = 0.9$. In panel (A), the points of equal mutation rates are indicated by solid circles for $\mu = 10^{-1}$, solid squares for both $\mu = 10^{-2}$ and $\mu = 2 \times 10^{-2}$, and solid triangles for $\mu = 3.16 \times 10^{-3}$ in the top right portion of the panel (corresponding to the range of mutation rate considered in panel (B) and in Figure 13, do not confuse triangles with arrowheads by μ). In panel (B) the same range of mutation is considered as in panel (B) of Figure 13. Average heritable organism fitness (expressed relative to defecting groups) is shown on the ordinate in panel (A) and is calculated as $\left(W_1 f_{11} x_1 + W_3(1 - x_1)\right)/W_3$ after the transition and $\left(W_2 f_{22} x_2 + W_4(1 - x_2)\right)/W_4$ before. Because the heritable fitness of cell groups is expressed relative to defecting groups, the heritable fitness of defecting groups is unity at all points in the figure.

in size than cooperating groups (but less functional).

Why do the modifiers evolve in this case of uniformly deleterious mutations; there is no longer any conflict between levels to mediate? Genotypes with the modifier have a lower level of deleterious mutational error. As discussed in section 4.8, the modifiers increase by virtue of increasing the heritability of fitness of the more fit non-mutant type. Both cells and groups are more fit, as a consequence of the fact that the modifier decreases the effective mutation rate. Organisms benefit twice from their lower mutation rate, because of their much larger size and enhanced functionality (panel (B) of Figure 14). The fitness of cells increases slightly as a result of the modifier although not as dramatically as at the level of the cell group.

The spectrum of fitness variation is similar for uniformly deleterious mutations (Figure 14) before and after the modifier increases, in dramatic contrast to selfish mutations (Figure 13). For example, a given decrease in the mutation rate (from 10^{-6} to 10^{-1}) affects the average heritable component of the fitness of the group by about the same amount (ordinate of Figure 14) whether the modifier is present or not, while the average fitness of cells has decreased to about 0.94 with the modifier compared to 0.90 without the modifier.

At all points along the curve in panels (A) of both Figure 13 and Figure 14, the heritable component of organism fitness is greater after the modifier increases than before. This result is shown clearly in panels (B) of both figures. The effect of the modifier on organism fitness is larger for selfish mutations than for uniformly deleterious mutations. There is one complication which is that when there is absolutely no mutation, the fitness at the organism level must be smaller after the modifier increases, because there is then no benefit of the modifier to offset the cost of allocating cells to the germ line. However, in the particular case considered in Figure 13 and Figure 14, the cost of the germ line is small because there are only five cell divisions in the germ line and so there is a relatively small number of cells in the germ line. This raises a limitation of the model, which is that the fitness of the organism (expected number of gametes produced) is assumed to scale with adult size and so to depend upon the number of cells in the soma, not the number of cells in the germ line. This may be reasonable, in many situations, but as the number of cells in the germ line decreases at some point the number of germ line cells must limit fitness. We have not included such an effect in our model to date, but plan to in the near future.

The basic consequence of an evolutionary transition is to move from a situation characterized by low between-group and high within-group change to the opposite situation characterized by high between-group and low within-group change. The fitness of defecting zygotes is not affected by the modifier, because there is no variation within D zygotes ($W_3 = W_4$). However, the heritability of fitness increases for cooperating zygotes. Concomitantly, the level of within-group change decreases as shown in Figure 12. As a result, after the transition, the fitness variation in Figure 13 is characterized by a steeper negative slope, indicating smaller variation within the cell group and greater variation between cell groups (organisms). After the transition the opportunity for within-group change is reduced while that for between-group change is enhanced.

The evolution of modifiers of within-organism change leads to increased levels of cooperation within the organism and increased heritability of fitness at the organism level. The evolution of these conflict mediators are the first new functions at the organism level. An organism is more than a group of cells related by common descent; organisms require adaptations that regulate conflict within. Otherwise their continued evolution is frustrated by the creation of within-organism variation and conflict. The evolution of modifiers of within-organism change are a necessary prerequisite to the emergence of individuality and the continued well being of the organism.

What happens during an evolutionary transition to a new higher level unit of individuality, in this case the multicellular organism? While taking fitness away from lower level units, cooperation increases the fitness of the new higher level unit. In this way, cooperation may create new higher levels of selection. However, the evolution of cooperation sets the stage for conflict, represented here by the increase of mutants within the emerging organism. The evolution of modifiers restricting within-organism change are the first higher level functions at the organism level. Before the evolution of a means to reduce conflict among cells, the evolution of new adaptations (such as the underlying traits leading to increased cooperation among cells) is frustrated by deleterious and defecting mutants. Individuality requires more than just cooperation among a group of genetically related cells, it also depends upon the emergence of higher level functions that restrict the opportunity for conflict within and ensure the continued cooperation of the lower level units. Conflict leads—through the evolution of adaptations that reduce it—to greater individuality and harmony for the organism.

References

1. Ameisen, J. C. 1996. The origin of programmed cell death. *Science (Washington, D.C.)* 272: 1278-1279.
2. Anderson, P. 1997. Kinase cascades regulating entry into apoptosis. *Microbiological Reviews* 61: 33-46.
3. Antolin, M. F. and Strobeck, C. 1985. The population genetics of somatic mutation in plants. *American Naturalist* 126: 52-62.
4. Bell, G. 1985. The origin and early evolution of germ cells as illustrated by the Volvocales. In *The Origin and Evolution of Sex*, ed. H. O. Halvorson, A. Monroy. 221-256. Alan R. Liss, Inc., New York.
5. Bell, G. and Koufopanou, V. 1991. The architecture of the life cycle in small organisms. *Philosophical Transactions of the Royal Society of London B, Biological Sciences* 332: 81-89.
6. Boraas, M. E., Seale, D. B., and Boxhorn, J. E. 1998. Phagotropphy by a flagellate selects for colonial prey: A possible origin of multicellularity. *Evolutionary Ecology* 12: 153-164.
7. Boyd, R. and Richerson, P. J. 1992. Punishment allows the evolution of cooperation (or anything else) in sizable groups. *Ethology and Sociobiology* 13: 171-195.
8. Buss, L. W. 1985. The uniqueness of the individual revisited. In *Population Biology and Evolution of Clonal Organisms*, ed. J. B. C. Jackson, L. W. Buss, R. E. Cook. Yale University Press, New Haven.
9. Buss, L. W. 1987. *The Evolution of Individuality*, Princeton, NJ: Princeton University.
10. Carson, D. A. and Ribeiro, J. M. 1993. Apoptosis and disease. *Lancet* 341: 1251-1254.
11. Cushing, J. M. 1989. A strong ergodic theorem for some nonlinear matrix models for the dynamics of structured populations. *National Resource Modeling* 3: 331-357.
12. Darwin, C. 1859. *The Origin of Species by Means of Natural Selection, or Preservation of Favoured Races in the Struggle for Life*, London: John Murray.
13. Drake, J. W. 1991. A constant rate of spontaneous mutation in DNA-based microbes. *Proceedings of the National Academy of Sciences, USA* 88: 7160-7164.
14. Fagerström, T. 1992. The meristem-meristem cycle as a basis for defining fitness in clonal plants. *Oikos* 63: 449-453.

15. Fagerström, T., Briscoe, D. A., and Sunnucks, P. 1998. Evolution of mitotic cell-lineages in multicellular organisms. *Trends in Ecology & Evolution* 13: 117-120.

16. Falconer, D. S. 1989. *Introduction to Quantitative Genetics*, Burnt Mill, England: Longman.

17. Frank, S. A. 1995. Mutual policing and repression of competition in the evolution of cooperative groups. *Nature (London)* 377: 520-522.

18. Gill, D. E., Chao, L., Perkins, S. L., and Wolf, J. B. 1995. Genetic mosaicism in plants and clonal animals. *Annual Review of Ecology and Systematics* 26: 423-444.

19. Gillott, M., Holen, D., Ekman, J., Harry, M., Boraas, M. E. 1993. Predation-induced *E. coli* filaments: Are they multicellular? In *Proceedings of the 51st Annual Meeting of the Microscopy Society of America*, ed. G. Baily, C. Reider. San Francisco Press, San Francisco, CA.

20. Godelle, B. and Reboud, X. 1995. Why are organelles uniparentally inherited. *Proceedings of the Royal Society of London B, Biological Sciences* 259: 27-33.

21. Godelle, B. and Reboud, X. 1997. The evolutionary dynamics of selfish replicators: a two-level selection model. *Journal of Theoretical Biology* 185: 401-413.

22. Haldane, J. B. S. 1937. The effect of variation on fitness. *American Naturalist* 71: 337-349.

23. Hopf, F., Michod, R. E., and Sanderson, M. J. 1988. The effect of reproductive system on mutation load. *Theoretical Population Biology* 33: 243-265.

24. Hull, D. L. 1981. Individuality and selection. *Annual Review of Ecology and Systematics* 11: 311-332.

25. Klekowski, E. J., Jr. 1988. *Mutation, Developmental Selection, and Plant Evolution*, New York: Columbia University Press.

26. Kondrashov, A. S. 1994. Mutation load under vegetative reproduction and cytoplasmic inheritance. *Genetics* 137: 311-318.

27. Leadbeater, B. S. C. 1983. Life-history and ultrastructure of a new marine species of *Proterospongia* (Choanoflagellida). *Journal of the Marine Biological Association of the United Kingdom* 63: 135-160.

28. Lewontin, R. C. 1970. The Units of Selection. *Annual Review of Ecology and Systematics* 1: 1-18.

29. Lewontin, R. C. 1978. Adaptation. *Scientific American* 239: 212-230.

30. Margulis, L. 1981. *Symbiosis in Cell Evolution*, San Francisco: W. H. Freeman.

31. Margulis, L. 1993. *Symbiosis in Cell Evolution, Microbial Communities in the Archean and Proterozoic Eons*, New York: W. H. Freeman.

32. Maynard Smith, J., Szathmáry, E. 1995. *The Major Transitions in Evolution*, San Francisco: W.H. Freeman.

33. Michod, R. E. 1980. Evolution of interactions in family structured populations: mixed mating models. *Genetics* 96: 275-296.

34. Michod, R. E. 1982. The theory of kin selection. *Annual Review of Ecology and Systematics* 13: 23-55.

35. Michod, R. E. 1995. *Eros and Evolution: A Natural Philosophy of Sex*, Reading, Mass.: Addison-Wesley.

36. Michod, R. E. 1996. Cooperation and conflict in the evolution of individuality. II. Conflict mediation. *Proceedings of the Royal Society of London B, Biological Sciences* 263: 813-822.

37. Michod, R. E. 1997. Cooperation and conflict in the evolution of individuality. I. Multi-level selection of the organism. *American Naturalist* 149: 607-645.

38. Michod, R. E. 1997. Evolution of the Individual. *American Naturalist* 150: S5-S21.

39. Michod, R. E. 1999. *Darwinian Dynamics, Evolutionary Transitions in Fitness and Individuality*, Princeton, N.J.: Princeton University Press.

40. Michod, R. E. and Roze, D. 1997. Transitions in individuality. *Proceedings of the Royal Society of London B, Biological Sciences* 264: 853-857.

41. Morrell, V. 1996. Genes versus teams. *Science (Washington, D.C.)* 273: 739-740.

42. Otto, S. P. and Orive, M. E. 1995. Evolutionary consequences of mutation and selection within an individual. *Genetics* 141: 1173-1187.

43. Price, G. R. 1970. Selection and covariance. *Nature (London)* 227: 529-531.

44. Price, G. R. 1972. Extension of covariance selection mathematics. *Annals of Human Genetics* 35: 485-490.

45. Price, G. R. 1995. The nature of selection. *Journal of Theoretical Biology* 175: 389-396.

46. Shikano, S., Luckinbill, L. S., and Kurihara, Y. 1990. Changes of traits in a bacterial population associated with protozoal predation. *Microbial Ecology* 20: 75-84.

47. Stanley, S. M. 1973. An ecological theory for the sudden origin of multicellular life in the Late Precambrian. *Proceedings of the National Academy of Sciences, USA* 70: 1486-1489.

48. Szathmáry, E. 1994. Toy models for simple forms of multicellularity, soma and germ. *Journal of Theoretical Biology* 169: 125-132.

49. Szathmáry, E. and Maynard Smith, J. 1995. The major evolutionary transitions. *Nature (London)* 374: 227-232.

50. Urushihara, H. 1992. Sexual development of cellular slime molds. *Development Growth & Differentiation* 34: 1-7.

51. Vogel, F. and Rathenberg, R. 1975. Spontaneous mutations in man. *Advances in Human Genetics* 5: 223-318.

52. Wade, M. J. 1985. Soft selection, hard selection, kin selection, and group selection. *American Naturalist* 125: 61-73.

53. Whitham, T. G. and Slobodchikoff, C. N. 1981. Evolution by individuals, plant-herbivore interactions, and mosaics of genetic variability: the adaptive significance of somatic mutations in plants. *Oecologia (Berlin)* 49: 287-292.

54. Wilson, E. O. 1975. *Sociobiology: The New Synthesis*, Cambridge, MA: Belknap Press of Harvard Univ. Press.

Appendix: equilibria for modifier model (with $G = 0$)

There are four dynamical equations (Equation 4) and the equilibria described in Table 6, and given mathematically in Table A 1, are obtained by setting the change in genotype frequencies equal to zero, along with $G = 0$ (Equation 3). The modifier allele may either be absent (Eq. 3) or fixed (Eq. 4) in the population, because there is no mutation affecting this allele.

There are four state variables corresponding to the frequencies of the four gamete types, but only three are independent, since they must sum to one. Consequently, there are three eigenvalues determining the stability of the system stemming from the Jacobian matrix of the linearized system evaluated at the different equilibria in Table A 2. Some of these eigenvalues depend on whether reproduction is sexual or asexual and the value of the recombination rate r. As seen in Table A 2, the eigenvalues are ratios of fitnesses multiplied by cell type frequencies (C or D). These eigenvalues can be expressed as ratios of fitnesses multiplied by heritability of fitness at the organism level. Equilibrium 2 (no cooperation, modifier allele fixed) is never stable, as we assume that the modifiers of within-organism change accrue some cost. Because of the cost of the modifier, $W_4 > W_3$, whatever the values of the parameters, and so $\lambda_{23} > 1$. This means that the evolution of functions to protect the integrity of the organism are not possible, if there is no conflict among the cells in the first place. It is conflict itself which sets the stage for the evolution of individuality.

Eq.	Genotype Frequencies	Allele Frequencies	Description of Loci	Interpretation
1	$x_1 = 0$, $x_2 = 0$, $x_3 = 0$, $x_4 = 1$	$q = 0$, $s = 0$	no cooperation; no modifier	*single cells, no organism*
2	$x_1 = 0$, $x_2 = 0$, $x_3 = 1$, $x_4 = 0$	$q = 0$, $s = 1$	no cooperation; modifier fixed	not of biological interest
3	$x_1 = 0$ $x_2 = \dfrac{W_2 \dfrac{k_{22}}{k_2} - W_4}{W_2 - W_4}$ $x_3 = 0$ $x_4 = \dfrac{k_{42}W_2}{k_2(W_2 - W_4)}$	$q = \dfrac{W_2 \dfrac{k_{22}}{k_2} - W_4}{W_2 - W_4}$ $s = 0$	polymorphic for cooperation and defection no modifier	*group of cooperating cells*: no higher level functions
4	$x_1 = \dfrac{W_1 \dfrac{K_{11}}{K_1} - W_3}{W_1 - W_3}$ $x_2 = 0$ $x_3 = \dfrac{K_{31}W_1}{K_1(W_1 - W_3)}$ $x_4 = 0$	$q = \dfrac{W_1 \dfrac{K_{11}}{K_1} - W_3}{W_1 - W_3}$ $s = 1$	polymorphic for cooperation and defection modifier fixed	*organism*: integrated group of cooperating cells with higher level function mediating within-organism conflict

Table A 1. Equilibria of Two Locus Modifier Model (Equation 4) with $G = 0$ (Equation 3). The variables q and s are the frequencies of the C and M alleles in the population.

The eigenvalues of the other three equilibria, in the case of asexual reproduction, indicate that two or three equilibria cannot be stable at the same time. In the case of sexual reproduction, this is no longer true. Although we have not been able to determine this analytically, because we have not obtained simple expressions for the eigenvalues ($\lambda_{31}, \lambda_{32}, \lambda_{41}$ and λ_{43} in Table A 2), as discussed in the text, using numerical experiments, we have discovered some regions in which equilibrium 1 and 4 are both stable and regions in which none of the four equilibria are stable. As discussed in the text, in the latter case there exist equilibria with $G > 0$.

	Equilibrium 1			Equilibrium 2		
	λ_{11}	λ_{12}	λ_{13}	λ_{21}	λ_{22}	λ_{23}
asexual	$\dfrac{W_1 \frac{K_{11}}{K_1}}{W_4}$	$\dfrac{W_2 \frac{k_{22}}{k_2}}{W_4}$	$\dfrac{W_3}{W_4}$	$\dfrac{W_1 \frac{K_{11}}{K_1}}{W_3}$	$\dfrac{W_2 \frac{k_{22}}{k_2}}{W_3}$	$\dfrac{W_4}{W_3}$
sexual	$\dfrac{(1-r)W_1 \frac{K_{11}}{K_1}}{W_4}$	$\dfrac{W_2 \frac{k_{22}}{k_2}}{W_4}$	$\dfrac{W_3}{W_4}$	$\dfrac{W_1 \frac{K_{11}}{K_1}}{W_3}$	$\dfrac{(1-r)W_2 \frac{k_{22}}{k_2}}{W_3}$	$\dfrac{W_4}{W_3}$
	Equilibrium 3			Equilibrium 4		
	λ_{31}	λ_{32}	λ_{33}	λ_{41}	λ_{42}	λ_{43}
asexual	$\dfrac{W_1 \frac{K_{11}}{K_1}}{W_2 \frac{k_{22}}{k_2}}$	$\dfrac{W_3}{W_2 \frac{k_{22}}{k_2}}$	$\dfrac{W_4}{W_2 \frac{k_{22}}{k_2}}$	$\dfrac{W_2 \frac{k_{22}}{k_2}}{W_1 \frac{K_{11}}{K_1}}$	$\dfrac{W_3}{W_1 \frac{K_{11}}{K_1}}$	$\dfrac{W_4}{W_1 \frac{K_{11}}{K_1}}$
sexual	$\dfrac{a_1 + \sqrt{b_1}}{c_1}$	$\dfrac{a_1 + \sqrt{b_1}}{c_1}$	$\dfrac{W_4}{W_2 \frac{k_{22}}{k_2}}$	$\dfrac{a_2 - \sqrt{b_2}}{c_2}$	$\dfrac{W_3}{W_1 \frac{K_{11}}{K_1}}$	$\dfrac{a_2 + \sqrt{b_2}}{c_2}$

Table A 2. Eigenvalues for asexual and sexual reproduction for increase of modifier. For each of the four equilibria described in Table A 1, the eigenvalues are given for both asexual (top) and sexual (reproduction). Although the equilibrium frequencies do not depend upon reproductive system, the stability does. The eigenvalues are ratios of products of group fitnesses and heritabilities [39]. For example, at equilibrium 3 the only genotypes in the population are genotypes 2 (*Cm*) and 4 (*Dm*). In the case of sexual reproduction a_1, b_1, c_1 and a_2, b_2, c_2 are complicated terms given in Table A 3.

$$a_1 = (1-r)K_{11}k_2 W_1 (W_2 - W_4)^2$$
$$+ K_1 \begin{pmatrix} r(W_1 - W_3)(W_2 - W_4)(k_{22}W_2 - k_2 W_4) \\ + k_2 W_3 (W_2 - W_4)^2 \end{pmatrix}$$

$$a_2 = K_{11}W_1 \begin{pmatrix} -rK_{11}k_2 W_1 W_2 - (1-r)k_{22}K_1 W_1 W_2 \\ +(1-r)k_{22}K_1 W_2 W_3 + rk_2 K_1 W_2 W_3 \\ +rK_{11}k_2 W_1 W_4 \\ -k_2 K_1 W_1 W_4 + (1-r)k_2 K_1 W_3 W_4 \end{pmatrix}$$

$$b_1 = -4K_{11}k_2^2 K_1 W_1 W_3 (1-r)(W_2 - W_4)^3 +$$
$$\begin{pmatrix} -(1-r)K_{11}k_2 W_1 (W_2 - W_4)^2 \\ + K_1 \begin{pmatrix} -r(W_1 - W_3)(W_2 - W_4)(k_{22}W_2 - k_2 W_4) \\ -k_2 W_3 (W_2 - W_4)^2 \end{pmatrix} \end{pmatrix}^2$$

$$b_2 = 4(r-1)k_{22}K_{11}^2 k_2 K_1^2 W_1^2 W_2 (W_3 - W_1)^2 W_4$$
$$+ K_{11}^2 W_1^2 \begin{pmatrix} rK_{11}k_2 W_1 W_2 + (1-r)k_{22}K_1 W_1 W_2 \\ -(1-r)k_{22}K_1 W_2 W_3 - rk_2 K_1 W_2 W_3 \\ -rK_{11}k_2 W_1 W_4 \\ +k_2 K_1 W_1 W_4 - (1-r)k_2 K_1 W_3 W_4 \end{pmatrix}^2$$

$$c_1 = 2k_{22}K_1 W_2 (W_2 - W_4)^2$$

$$c_2 = 2K_{11}^2 k_2 W_1^2 (W_3 - W_1)$$

Table A 3. Terms for eigenvalues under sexual reproduction given in Table A 2.

Richard E. Michod is in the Department of Ecology and Evolutionary Biology University of Arizona, Tucson, AZ 85721, email address: michod@u.arizona.edu.

Denis Roze is in the Department of Ecology and Evolutionary Biology University of Arizona, Tucson, AZ 85721, email address: roze@u.arizona.edu.

Lectures on Mathematics in the Life Sciences
Volume **26**, 1999

On the Manner in which Biological Complexity May Grow

Chrystopher L. Nehaniv and John L. Rhodes

ABSTRACT. Previously in (*Mathematical Hierarchies & Biology*, AMS Series in Discrete Math. & Theoretical Computer Science, vol. 37, 29–42, 1997) the authors gave an algebraic definition of the complexity of biological systems, exhibited the unique maximal integer-valued complexity measure satisfying it, and proved bounds on evolutionary jumps in complexity: If complexity is N, it can under a single smooth transition jump at most to $2N + 1$. We review these results and address the question of, *How fast can complexity evolve over longer periods of time?* Although complexity may more than double in a single generation, we prove that in a smooth sequence of t 'inclusion' steps, complexity may grow at most from N to $(N + 1)t + N$, a linear function of number of generations t. While for sequences of 'mapping' steps it increases by at most t. Thus, despite the fact that there are major transitions in which complexity jumps are possible, over longer periods of time, the growth of complexity may be broken into maximal intervals on which it is bounded in the manner described.

1. Complexity Measures for Living Organisms

Attempts to measure the complexity of organisms in a systematic fashion have resulted in several proposals, surveyed by Bonner [**Bon88**, Ch. 6], Raff and Kaufman [**RK83**, Ch. 11], and Nehaniv and Rhodes [**NR97**]. These include *genome size* (the so-called C-value), the number of base-pairs of DNA in the haploid complement of an organism's genome. Unfortunately, C-value varies widely even for closely related taxa of apparently similar complexity. The *number of cell types* [**Bon65, Kau69**] a multicellular organism may have during the course of its life cycle is a better complexity measure, and Kauffman has argued that the number of cell types corresponds to the number of stable states for a given genome, relating this measure to an automata-theoretic viewpoint.

Our approach is to take the informal notion *number of levels in a hierarchy of organization* and axiomatize it to obtain a complexity measure for biological systems modelled as finite automata [**NR97**]. This complexity measure is well-defined, and has the virtue of mathematical precision. While its particular values on automata models of a biological system may be difficult to calculate or adequate models may not be available until further biological details unfold, the value of the complexity

---

1991 *Mathematics Subject Classification*. Primary 92D15, 20M20, 92B05; Secondary 20M35, 68Q70.

measure on each particular automata model of any particular biological system has
an unchanging value. Furthermore, our hierarchical complexity measure is related
by numerical dominance to all others satisfying the complexity axioms as the unique
measure having this maximality property. Taking an algebraic perspective, we seek
insight into and predictions for understanding the evolution of biological systems.

2. Mathematical and Notational Preliminaries

We lay the algebraic groundwork for constructing the unique maximal complexity measure on biological systems modelled as finite automata.

Semigroups, Automata, Transformation Semigroups A *semigroup* is a set S
with a binary multiplication defined on it satisfying the associative law: $(xy)z =
x(yz)$. An *automaton* $X = (Q, S)$ is a set of states Q together with, for each symbol
s of an input alphabet S, an associated [state-transition] function from Q to Q (also
denoted by s). A special case of this is a *transformation semigroup* (Q, S), which
is a right action of a semigroup S on a set of states Q, *i.e.*, a rule taking $q \in Q$ and
$s \in S$ to $q \cdot s \in Q$, satisfying $(q \cdot s) \cdot s' = q \cdot ss'$ for all $q \in Q$, and $s, s' \in S$. Given any
automaton X we canonically associate to it a transformation semigroup generated
by considering the semigroup of state-transition mappings induced by all possible
sequences of inputs to the automaton acting on its states. To avoid trivialities,
we shall usually assume that state sets, input sets, and semigroups are non-empty.
Generally we shall require that two distinct inputs $a \neq a'$ to an automaton differ
at least at one state x: $x \cdot a \neq x \cdot a'$. Such an automaton is called *faithful*. The
restriction of faithfulness is not essential, for one can 'faithfulize' X by identifying
input symbols which have the same action.

For semigroups, a *homomorphism* (or, more briefly, a *morphism*) from S to T
is a function $\varphi : S \to T$ such that $\varphi(ss') = \varphi(s)\varphi(s')$ for all $s, s' \in S$. A morphism
known to be surjective (onto) will be called a *surmorphism* and be denoted with a
double-headed arrow $\varphi : S \twoheadrightarrow T$; the semigroup T is then said to be a *homomorphic
image* of S. An injective (one-to-one) morphism φ is called an *embedding* of S into
T. T is a *subsemigroup* of S if T is a subset of S closed under multiplication. We
write $T \leq S$ if T is a subsemigroup of S or, more generally, embeds in S. If there
is a bijective (one-to-one and onto) morphism from S to T, we say that S and T
are *isomorphic* and write $S \cong T$. The notation $S \prec T$ means S is a homomorphic
image of a subsemigroup of T. Intuitively, T can emulate any computation of S.
In such a case, we say S *divides* T. Division is a transitive relation.

If $X = (Q, S)$ and $Y = (U, T)$ are automata or transformation semigroups a
morphism φ from X to Y consists a function $\varphi_1 : Q \to U$ and a mapping $\varphi_2 : S \to T$
(which is required to a semigroup morphism in the transformation semigroup case),
such that for all $q \in Q$ and $s \in S$, one has $\varphi_1(q \cdot s) = \varphi_1(q) \cdot \varphi_2(s)$. Surmorphism,
embedding, isomorphism and division of transformation semigroups are defined
analogously. If Y divides X, it is also common to say that X *covers* Y and that X
emulates Y.

Given a semigroup S, we define $S^\bullet$ to be S if S contains an identity element
$1 \in S$ with $s1 = 1s = s$ for all $s \in S$, otherwise we take $S^\bullet$ to be S with new
identity element 1 adjoined. If $X = (Q, S)$ is an automaton or transformation
semigroup, we define $X^\bullet = (Q, S^\bullet)$, where the identity of $S^\bullet$ acts as the identity
function on all states in Q. Also $\overline{X} = (Q, \overline{S})$, where for each state $q \in Q$ a constant

map taking value q has been adjoined as an element of S. For a semigroup S, one obtains a canonically its so-called *right regular representation* $(S^\bullet, S)$, with states $S^\bullet$, semigroup S and action $s \cdot s' = ss'$ for all $s \in S^\bullet$ and $s' \in S$.

Ideals, Groups, and Green's Relations in Semigroups. A subsemigroup I of S is called an *ideal* if $SIS \subseteq I$, and a *left ideal* if $SI \subseteq I$. *Right ideals* are defined analogously. One says that for $s, t \in S$, that $s \geq_J t$ (read: "s is $\mathcal{J}$-*above*" t) if $t \in S^\bullet s S^\bullet$, the principal ideal generated by s. This comprises a transitive relation on S. One says that t is $\mathcal{J}$-*equivalent* to s if s and t generate the same principal ideals. A $\mathcal{J}$-*class* of S is maximal subset of S consisting of $\mathcal{J}$-equivalent elements. The notation $s >_J t$ (read: "s is strictly $\mathcal{J}$-above t") means that s is $\mathcal{J}$-above but not $\mathcal{J}$-equivalent t, while $s \not>_J t$ is the negation of this. A $\mathcal{J}$-class is said to be *null* if $ss' \notin J$ for all $s, s' \in J$. An element $s \in S$ is *regular* if its $\mathcal{J}$-class is not null.

By considering left principal ideals, one obtains analogously the transitive $\mathcal{L}$-*relation*: $s \geq_L t$ (read: "s is $\mathcal{L}$-above t") if $t \in S^\bullet s$. So one has also notions of $\mathcal{L}$-*equivalence* and $\mathcal{L}$-*class*. The $\mathcal{R}$-*relation* $\geq_R$ may be defined as the dual of the $\mathcal{L}$-relation.

A *group* is a semigroup S with exactly one $\mathcal{L}$-class and exactly one $\mathcal{R}$-class; or, equivalently, S has an identity e and for each element s in S, there exists an *inverse* s' in S with $ss' = s's = e$. A group is called *simple* if its homomorphic images are just itself and the one element group (up to isomorphism). A finite semigroup each of whose subgroups have only one element is called *aperiodic*.

Wreath Products. Given two transformation semigroups (X, S) and (Y, T), we define the *wreath product* $(Y, T) \wr (X, S)$ to be the transformation semigroup with set $Y \times X$ and action semigroup consisting of pairs (f, s) with $f \in T^X$ and $s \in S$ with action $(y, x) \cdot (f, s) = (y \cdot f(x), x \cdot s)$. This defines a transformation semigroup with the evident multiplicative structure on its action semigroup. Moreover the wreath operation on the class of transformation semigroups is associative. *If S and T are semigroups, we shall write $S \wr T$ for the wreath product of their right regular representations.* This corresponds to generic hierarchical arrangement.

3. Axiomatizing Hierarchical Complexity

We axiomatize the notion of hierarchical complexity of automata models of biological systems, and then study the implications of the algebraic approach for the mathematical analysis of hierarchical complexity and its evolution.

In our axiomatization of hierarchical complexity, we will not be concerned with how a biological system is built, or even about the 'right' components to identify or the 'right' boundaries to draw between its levels. Generally one expects that endeavoring to answer the above will require one to frame difficult and important questions of research for biochemistry, molecular genetics, cellular biology, and physiology, anatomy, development and ecology of particular organisms.

Fortunately, by asking only for the hierarchical complexity and not for an explicit hierarchical description, the above issues can be initially avoided and addressed as the state of our biological knowledge expands. Assuming the automata-theoretic modelling of biological systems can be done in principle is all that is needed.

Given any automaton X, we canonically associate to it a transformation semi-group generated by considering the semigroup of state-transition mappings induced by all possible sequences of inputs to the automaton acting on its states.

3.1. Complexity Axioms. A function $c : \mathcal{A} \to \mathbb{N}$, from transformation semi-groups canonically associated to finite automata (modelling organisms) to the natural numbers is called a *hierarchical complexity measure* if it satisfies:

1. (**Bounded Emergence**) *If X and Y are combined hierarchically, the result has complexity not exceeding the sum of complexities of X and Y:*

$$c(X \wr Y) \leq c(X) + c(Y)$$

2. (**Non-Interaction**) *If X and Y are combined, but do not interact, the complexity of the whole is just the maximum of complexities of X and Y:*

$$c(X \times Y) = \max(c(X), c(Y))$$

3. (**Covering**) *If Y can do everything that X can, then X is not more complex than Y:*

$$\text{If } X \prec Y, \text{ then } c(X) \leq c(Y)$$

4. (**Constructibility**) *Every X can be constructed from components of minimal complexity:*

$$\text{For all } X, \ X \prec X_n \wr \cdots \wr X_1, \text{ for some } X_i \text{ with } c(X_i) \leq 1$$

5. (**Initial Condition**) *If X has only trivial subgroups then $c(X) = 0$.*

Axiom 1 states that although complexity can emerge through interaction how much can emerge is a bounded function of the complexity of the interacting components.[1] Axiom 2 says there can be no complexity increase without interaction.[2] Axiom 3 says a part cannot be more complex than the whole. Axioms 1, 2, 3 should obviously be properties of any good complexity measure. Axiom 4 reflects the scientific optimism that everything should be constructible from fundamental building blocks, i.e. components of minimal complexity. Axiom 5 says that (at least) certain particularly elementary building blocks –those which cannot count modulo n for any integer $n \geq 2$, should have zero complexity.

The above axioms can be satisfied by assigning to every model of each organism the value zero. So complexity measures exist. If, in addition, a complexity measure c satisfies the following axiom, then it is called *the [maximal] complexity measure*.

6. (**Maximality**) *If $f : \mathcal{A} \to \mathbb{N}$ satisfies the preceding axioms, then for all X, we have $f(X) \leq c(X)$.*

---

[1]This axiom does *not* imply that whole is just the sum of the parts. On the contrary, it implies that complexity can emerge through interaction (and, by axiom 2, cannot emerge without it). It does not however allow unbounded emergence, but rather that complexity is constrained by the components and their interaction. The reason for choosing addition as the bounding function is its mathematical convenience and tractability.

[2]The complexity of a whole is just the maximum complexity of its parts if they don't interact.

4. Construction of the Complexity Measure

We now show that existence and uniqueness of the hierarchical complexity measure follows from the Krohn-Rhodes Prime Decomposition Theorem [**KR65**].

THEOREM 4.1 (**Prime Decomposition Theorem**). *For every finite transformation semigroup X, there exist finite groups and aperiodic semigroups X_i such that*

$$X \prec X_n \wr \cdots \wr X_1. \qquad (\star)$$

(One can build any X from group and aperiodic semigroup automata.) Moreover, if G is a finite simple group which divides the semigroup of X, then G divides the semigroup of X_i for some $1 \le i \le n$. Moreover, this fact about finite simple groups is true for every possible decomposition.

LEMMA 4.2. *If $c : \mathcal{A} \to \mathbb{N}$ is any complexity function, and G is a group, then $c(G) \le 1$.*

Proof: First we consider the case of a simple group G. By axiom 4, $G \prec X_1 \wr \ldots \wr X_n$ with $c(X_i) \le 1$ for all i. But by the Prime Decomposition Theorem, it follows that $G \prec X_i$ for some i, whence by axiom 3, $c(G) \le c(X_i) \le 1$. Now consider any non-trivial group G. It is easy to see that G is isomorphic to a subgroup of the alternating group A_{2n} where $n = |G|$, which is a simple group. Thus, by axiom 3, it follows that $c(G) \le c(A_{2n}) \le 1$. $\qquad \square$

THEOREM 4.3. *There exists a unique maximal complexity measure $cpx : \mathcal{A} \to \mathbb{N}$. To define $cpx(X)$, consider all decompositions of the form $(\star)$. We have :*

$$X \prec A_n \wr G_n \wr A_{n-1} \wr \cdots \wr A_i \wr G_i \wr A_{i-1} \wr \cdots \wr A_1 \wr G_1 \wr A_0,$$

where A_i has no nontrivial subgroup and G_i is a nontrivial group.
Then $cpx(X) = $ least n for which such a decomposition of X exists.

Note: *$cpx(X)$ is the least number of active computing (group) levels required in order to build X hierarchically from simple components.*

Proof: Axioms 4 and 5 are immediate for cpx, and 3 follows from transitivity of '$\prec$'. Since the wreath product of groups is a group and the wreath product of aperiodics is aperiodic, axiom 1 follows by hierarchically combining minimal length decompositions of X and Y. By taking direct products of components in shortest decompositions of X and Y, one obtains axiom 2. Finally, if c is a complexity measure satisfying axioms 1-5, then taking a shortest Krohn-Rhodes decomposition of X as above:

$$\begin{aligned} c(X) &\le & c(A_n \wr G_n \wr A_{n-1} \wr \cdots \wr A_1 \wr G_1 \wr A_0) \text{ by axiom 3} \\ &\le & c(A_n) + c(G_n) + \cdots + c(G_1) + c(A_0) \text{ by axiom 1} \\ &\le & n \text{ by axiom 5 and Lemma 4.2} \\ &\le & cpx(X), \end{aligned}$$

so axiom 6 holds. Moreover, the maximal complexity measure is unique, for, if cpx' were another, then $cpx'(X) \le cpx(X) \le cpx'(X)$ by two applications of axiom 6. $\qquad \square$

5. Evolutionary Transitions in Complexity

J. Maynard Smith and E. Szathmáry [**MSS95**] and L. Buss [**Bus87**] have identified major transitions in evolution: from self-replicating molecules to populations of molecules in bounded compartments; from independent replicators to chromosomes; from RNA as gene and enzyme to DNA and protein (advant of the genetic code); from prokaryotes to eukaryotes; from asexual clones to sexual populations; from protists to animals, plants, fungi (multicellularity, cell differentiation); from solitary individuals to colonies (with non-reproductive castes); from primate societies to human society (language).

A main source of major transitions of evolution [**MSS95**] and those in the evolution of individuality [**Bus87**] on our planet involves change in the manner in which information is represented or interpreted. This suggests that the application of automata theory to the study of evolutionary transitions is appropriate. It is worth noting that the current models of genomic control over cell state in cellular and molecular biology are essentially automata models, as already made explicit by the early mathematical biology work of Kauffman [**Kau72**] for feedback control of protein biosynthesis; by Krohn, Langer, and Rhodes [**KLR67**] for the Krebs cycle; and by Lindenmayer [**Lin68**] for simple models of development. A second major source of evolutionary transitions has been the appearance of new levels of hierarchical structure, as is obvious in the many examples discussed by [**MSS95**], and may involve processes from which originate new units of fitness on which natural selection and evolution act [**MR**].

Assuming descent is "as smooth as can be" — without 'hopeful monsters" and moreover without any intermediates being skipped over — in the evolution of hierarchical complexity,[3] we ask, *How does complexity increase, smoothly or with jumps?* We can study this question for single steps in evolutionary change (section 5.1 or [**NR97**]) or over longer periods of time (section 5.2).

Formally, suppose that during descent with modification we have a sequence of organisms (represented by the semigroups associated to the finite-state automata modelling them) of increasing computational power:

$$S_0 \prec S_1 \prec S_2 \prec \cdots \prec S_n.$$

Suppose, furthermore, that no proper intermediates could possibly be introduced into the sequence:

$$\text{If } S_i \prec T \prec S_{i+1}, \text{ then } S_i \cong T \text{ or } T \cong S_{i+1}.$$

Each division $S_i \prec S_{i+1}$ is "maximal". A division $S_i \prec S_{i+1}$ is called *proper* if $S_i \neq S_{i+1}$. We will study the hierarchical complexity changes possible in a single step of such an unrefinable chain.[4]

Whether or not a particular chain of semigroups in such a sequence can arise as the the chain associated to a sequence of biological systems related by descent, of course, will depend on biological factors.

---

[3]Without the assumption of smooth evolution or other assumptions, the size of evolutionary jumps cannot be bounded. The considerations here study the complexity jumps and the manner of complexity growth over longer periods when smoothness is assumed.

[4]One might also consider unrefinable chains with each step either increasing or decreasing in computational power. Since this just corresponds to the reversal of some maximal divisions '$\prec$', our results — which bound the complexity changes on the two sides of the '$\prec$' – will apply also to this more general situation.

5.1. Bounds on Single Complexity Jumps in Evolution.

THEOREM 5.1 (**Jump Lemma [NR97]**). *Let $S \prec T$ be a maximal proper division of finite semigroups. Then we have either:*
(1) (**mapping step**) S *is a homomorphic image of T and*

$$cpx(S) \le cpx(T) \le cpx(S) + 1,$$

or
(2) (**inclusion step**) S *is a maximal subsemigroup of T and*

$$cpx(S) \le cpx(T) \le 2\, cpx(S) + 1.$$

Proof: By hierarchical complexity axiom 3, one has that $cpx(S) \le cpx(T)$. By definition of division ($\prec$), we have that $S \leftarrow\!\!\!\leftarrow T' \le T$ for some subsemigroup T' of T mapping homomorphically onto S. By maximality of the division, it follows that either (1) $T = T'$ (so $S \leftarrow\!\!\!\leftarrow T$ is a maximal proper surmorphism [**Rho67, KRT68**]) or (2) $S \cong T'$ (so S is a maximal proper subsemigroup of T [**GGR68, KRT68**]).

For case (1), by the classification of maximal proper surmorphisms ([**Rho67**], [**KRT68**] or [**RW89**]), the surjective morphism φ is either (i) injective when restricted to each subgroup of T, hence by the Fundamental Lemma of Complexity [**Rho68, Rho71b, Rho74**] or [**Til76**, Ch. XII], $cpx(S) = cpx(T)$; or (ii) for regular elements $t_1, t_2 \in T$, if $\varphi(t_1) = \varphi(t_2)$ then t_1 and t_2 are $\mathcal{L}$-equivalent, hence by [**AHNR95**], [**KRT68**], or [**RW89**], $cpx(T) \le cpx(S) + 1$. Case (2) will be handled in Theorem 5.3 by taking $t = 1$. ☐

Sharpness of the Bounds: 'Genius Jumps'. In fact, it is possible to construct $R \le T$, a maximal proper subsemigroup as above with $cpx(T) = 2\, cpx(R) + 1$, so these bounds are attained. Such a example construction is sketched in [**NR97**] and full details will be published elsewhere [**NR**]. Despite the more than doubling of complexity in such examples, it is not possible to have such increases in each step of chain (see Theorem 5.3 below).

The above 'genius jumps' rely on existing fragmentation / inefficiency or they would not be possible. An important factor in most major evolutionary transitions [**Ohn70, Bus87, MSS95**] is the hierarchical structuring of pre-existing units (with adaptations): division of labor / specialization / differentiation, duplication and divergence, symbiosis, epigenesis. The algebraic structure of known semigroup examples of 'genius jumps' seems conceptually close to transitions of this kind, and such structure may be typical of many evolutionary complexity jumps (cf. [**NR97**]). Whether and how shifts in the way biological systems use information, can be understood in this manner is unknown.

5.2. Bounds on Complexity Growth for Longer Term Change.
Although we have seen that short term change may involve a complexity jump from N to $2N + 1$ in a single step, such jumps cannot happen continually.

LEMMA 5.2. *Let I is an ideal of a semigroup with $\mathcal{J}$-class J, maximal in I. Denoting the ideal $I \setminus J$ by $\widehat{I}$, we have*

$$cpx(I) \le cpx(\widehat{I}) + 1.$$

Proof: We recall the '$\mathcal{V}$-union-T' technique [**KRT68**, Ch. 5, Sec. 4], [**Eil76**, Prop. 3.5] or [**Neh95**, Sec. 3]: Given a semigroup $S = V \cup T$ with V a left ideal and T a subsemigroup, we have $S \prec \overline{V}^{\bullet} \wr \overline{T}^{\bullet}$, and so $cpx(S) \le cpx(V) + cpx(T)$.

We also recall the Reduction Theorem ([**RT75**], [**Til76**, Ch. XII]): Let E be a system of idempotents[5] for S, i.e. one idempotent from each maximal non-null $\mathcal{J}$-class, then $cpx(ESE) = cpx(S)$. Taking E to be a system of idempotents for I,

$$EIE = E(\widehat{I} \cup J)E = E\widehat{I}E \cup EJE \subseteq \widehat{I} \cup G_e,$$

where G_e is the maximal subgroup of J containing an $e \in E$, or empty if J is a null $\mathcal{J}$-class. Now $\widehat{I} \cup G_e$ is a semigroup, so by 'V-union-T', one has: $EIE \prec \widehat{I}^{\bullet} \wr \overline{G_e}^{\bullet}$. So by the Reduction Theorem and Lemma 4.2, $cpx(I) = cpx(EIE) \leq cpx(\widehat{I}) + 1$. $\square$

THEOREM 5.3. *Consider a sequence of maximal proper inclusions:*

$$R = S_0 \subset S_1 \subset S_2 \subset \ldots \subset S_t = T$$

Then $cpx(T) \leq (cpx(R) + 1)t + cpx(R)$.

Remark: This theorem implies that for a chain of 'inclusion steps' in evolution, complexity is at most a linear function of time t with slope given by 1 plus the complexity of the first biological system in the chain.

Proof: We have S_i is a maximal subsemigroup of S_{i+1} for each $i = 0, \ldots, t - 1$. By [**GGR68**] or [**KRT68**], $S_{i+1} \setminus S_i$ is contained in some $\mathcal{J}$-class J_{i+1} of S_{i+1} Each of these $\mathcal{J}$-classes is contained in a unique $\mathcal{J}$-class of T (some of these $\mathcal{J}$-classes may coincide). Denote these distinct $\mathcal{J}$-classes of T by $J_1', \ldots, J_m'$ ($m \leq t$). Furthermore, we may re-index so that $J_i' >_{\mathcal{J}} J_j'$ implies $i > j$ ($1 \leq i, j \leq m$).

Now define for $k = 1, \ldots, m$, the following ideals of T:

$$I_k = \{t \in T : t \leq_{\mathcal{J}} J_i' \text{ some } i \leq k\}.$$

We have $i > j$ implies $I_j \subset I_i$. Let $\widehat{I}_k$ be the ideal $I_k \setminus J_k'$ of T.

Observe that

$$\widehat{I}_k \subseteq R \cup I_{k-1},$$

and that $R \cup I_{k-1}$ is a subsemigroup of T, with ideal I_{k-1} and subsemigroup R. By the '$V \cup T$'-construction, we have $cpx(R \cup I_{k-1}) \prec cpx(I_{k-1}) + cpx(R)$. Therefore,

$$cpx(\widehat{I}_k) \leq cpx(I_{k-1}) + cpx(R).$$

Furthermore, by Lemma 5.2, $cpx(I_k) \leq cpx(\widehat{I}_k) + 1$. So we can conclude,

$$cpx(I_k) \leq cpx(I_{k-1}) + cpx(R) + 1.$$

Then, inductively, we have that

$$cpx(I_m) \leq cpx(I_1) + (m - 1)(cpx(R) + 1).$$

Now, since $\widehat{I}_1 \subseteq R$, by Lemma 5.2, $cpx(I_1) \leq cpx(\widehat{I}_1) + 1 \leq cpx(R) + 1$. Whence,

$$cpx(I_m) \leq m(cpx(R) + 1).$$

But T is the union of ideal I_m and subsemigroup R, so using 'V-union-T':

$$cpx(T) \leq cpx(I_m) + cpx(R) \leq (m + 1)cpx(R) + m.$$

As a result, $cpx(T) \leq (t + 1)cpx(R) + t = (cpx(R) + 1)t + cpx(R)$. $\square$

---

[5]An idempotent is an element e such that $e^2 = e$.

We considered a sequence of finite semigroups

$$S_0 \prec S_1 \prec \cdots \prec S_t$$

representing an as continuous as possible sequence of biological systems.

Now at a 'mapping step' we saw in the Jump Lemma that complexity increases by at most 1, so then the associated increase in complexity in this chain is at most piecewise linear in time in the following sense: Take S_i and S_j $(i < j)$ in the sequence above so that all the maximal proper divisions in between them are of the same kind (i.e. either all of mapping type or all of inclusion type). In the mapping case, complexity grows by at most one per generation. In the inclusion case, it grows at most linearly as described in Theorem 5.3. Thus the graph of complexity increase is bounded above by linear sections of slope 1 and slope $cpx(S_i) + 1$, respectively, where i is the index of the first in a run of inclusions.

References

[AHNR95] B. Austin, K. Henckell, C. Nehaniv and J. Rhodes, Complexity and Subsemigroups via the Presentation Lemma, *Journal of Pure & Applied Algebra*, Vol. 101, No. 3, pp. 245–289, 1995.

[Bon65] J. Tyler Bonner, *Size and Cycle*, Princeton University Press, 1965.

[Bon88] J. Tyler Bonner. *The Evolution of Complexity by Means of Natural Selection*, Princeton, 1988.

[Bus87] L. W. Buss. *The Evolution of Individuality*, Princeton, 1987.

[Eil76] S. Eilenberg. *Automata, Languages and Machines*, volume B. Academic Press, New York, 1976.

[GGR68] N. Graham, R. Graham, and J. Rhodes, Maximal subsemigroups of finite semigroups, *Journal of Combinatorial Theory*, Vol. 4, No. 3, pp. 203–209, 1968.

[Kau69] S. A. Kauffman, Metabolic stability and epigenesis in randomly constructed genetic nets. *Journal of Theoretical Biology*, **22**, 437–467, 1969.

[Kau72] S. A. Kauffman, Organization of Cellular Genetic Control Systems. In J. D. Cowan, ed., *Some Mathematical Questions in Biology II*, Lectures on Mathematics in the Life Sciences, Vol. 3, American Mathematical Society, pp. 62–116, 1972.

[KR65] K. Krohn and J. Rhodes, Algebraic theory of machines, I. Prime decomposition theorem for finite semigroups and machines, *Transactions of the American Mathematical Society*, **116**, 450–464, 1965.

[KLR67] K. Krohn, R. Langer and J. Rhodes, Algebraic Principles for the Analysis of a Biochemical System. *Journal of Computer & Systems Sciences*, **1**, 119–136, 1967.

[KRT68] K. B. Krohn, J. L. Rhodes and B. R. Tilson. The Prime Decomposition Theorem of the Algebraic Theory of Machines (Chapter 5), Local Structure Theorems for Finite Semigroups (Chapter 7), and Homomorphisms and Semilocal Theory (Chapter 8). In M. Arbib, ed., *Algebraic Theory of Machines, Languages, and Semigroups*, Academic Press, 1968.

[Lin68] A. Lindenmayer, Mathematical Models for Cellular Interactions in Development: I. Filaments with One-sided Inputs, & II. Simple and Branching Filaments with Two-sided Inputs, *Journal of Theoretical Biology*, **18**, 290–299 and 300–315, 1968.

[MSS95] J. Maynard Smith and E. Szathmáry. *The Major Transitions in Evolution*, W. H. Freeman, 1995.

[MR] R. E. Michod and D. Roze, Cooperation and Conflict in the Evolution of Individuality. III. Transitions in the Unit of Fitness. In: *Mathematical & Computational Biology*, American Mathematical Society, (this volume).

[Neh95] C. L. Nehaniv, Cascade decomposition of arbitrary semigroups. In J. Fountain, ed., *Semigroups, Formal Languages and Groups*, (NATO Advanced Study Institute, University of York, 7-21 August 1993), Kluwer Academic Publishers, pp. 391–425, 1995.

[NR97] C. L. Nehaniv and J. L. Rhodes. Krohn-Rhodes Theory, Hierarchies, and Evolution. In: B. Mirkin, F.R. McMorris, F.S. Roberts, and A. Rzhetsky, eds., *Mathematical Hierarchies and Biology*. DIMACS Series in Discrete Mathematics and Theoretical Computer Science. Amer. Math. Society, Providence, pp. 29–42, 1997.

[NR]   C. L. Nehaniv and J. L. Rhodes, The evolution of biological complexity from an algebraic perspective, *in preparation.*

[Ohn70]   S. Ohno, *Evolution by Gene Duplication.* Springer-Verlag, 1970.

[RK83]   R. A. Raff and T. C. Kaufman. *Embryos, Genes, and Evolution: The Developmental-Genetic Basis of Evolutionary Change.* Macmillan Publishing Co., 1983. Reprinted with new introduction and preface, Indiana University Press, 1991.

[Rho67]   J. Rhodes, A homomorphism theorem for finite semigroups, *Mathematical Systems Theory,* **1**, 289–304, 1967.

[Rho68]   J. Rhodes, The fundamental lemma of complexity for arbitrary finite semigroups, *Bulletin of the American Mathematical Society,* **68**, 1104–1109, 1968.

[Rho71a] J. Rhodes, *Applications of Automata Theory and Algebra via the Mathematical Theory of Complexity to Biology, Physics, Psychology, Philosophy, Games, and Codes.* University of California Library (unpublished book, 1971); revised edition forthcoming.

[Rho71b] J. Rhodes, Proof of the fundamental lemma of complexity (weak version) for arbitrary finite semigroups, *Journal of Combinatorial Theory, Series A,* **10**, 22–73, 1971.

[Rho74]   J. Rhodes, Proof of the fundamental lemma of complexity (strong version) for arbitrary finite semigroups, *Journal of Combinatorial Theory, Series A,* Vol. 16, No. 2, pp. 209–214, 1974.

[RT75]   J. Rhodes and B. Tilson, A reduction theorem for complexity of finite semigroups, *Semigroup Forum,* **10**, 96–114, 1975.

[RW89]   J. Rhodes and P. Weil, Decomposition techniques for finite semigroups, using categories I & II, *Journal of Pure & Applied Algebra,* **62**, 269–284 and 285–312, 1989.

[Til76]   B. Tilson, Depth Decomposition Theorem (Chapter XI) & Complexity of Semigroups and Morphisms (Chapter XII). In S. Eilenberg, *Automata, Languages, and Machines,* Vol. B, Academic Press, 1976.

CYBERNETICS AND SOFTWARE SYSTEMS GROUP, UNIVERSITY OF AIZU, AIZU-WAKAMATSU CITY, FUKUSHIMA PREF. 965-8580, JAPAN

Current address: Interactive Systems Engineering, Department of Computer Science, University of Hertfordshire, Hatfield, Hertfordshire AL10 9AB, United Kingdom

E-mail address: `nehaniv@u-aizu.ac.jp, c.l.nehaniv@herts.ac.uk`

DEPARTMENT OF MATHEMATICS, UNIVERSITY OF CALIFORNIA, BERKELEY, CALIFORNIA 94720, U.S.A.

E-mail address: `rhodes@math.berkeley.edu`

Lectures on Mathematics in the Life Sciences
Volume **26**, 1999

A Simple Evolvable Development System in Euclidean Space

Tatsuo Unemi

ABSTRACT. This paper presents one alternative trial to simulate an evolutionary process of a developmental system for multi-cellular organisms. A simplified growth model of multi-cellular plants in two-dimensional (2D) and three-dimensional (3D) Euclidean space is proposed where the size and shape of each cell are fixed. Rules of growth are encoded as genes on a chromosome in each cell, and the population evolves through a Genetic Algorithm. Computer simulation of three different cases are described, with pre-defined fitness functions in 2D and 3D space, and an ecological fitness function on 2D space. The results demonstrate how simple genetic information can result in emergence of a wide diversity in phenotypic forms.

1. Introduction

It is one of the plausible views that the growth of multi-cellular plants is realized by iteration of cells' division, cohesion, enlargement, reformation, and death. These activities are triggered by some chemical and physical events on the cell itself guided by the genetic information on the chromosomes it contains. Through a lot of efforts by biologists, some details of species-specific developmental processes have been revealed, and it seems fully complicated and surprising as there are a very wide variety of strategies of morphogenesis. To deepen our understanding on what life is, it is also important to build mathematical models of biological activities in some abstraction, concurrently investigating more detail of concrete organisms.

One of the great tools for mathematical modeling of growth of multi-cellular plants is *L-systems* [**LI**], which provide a formal method with a type of *rewriting rule set* to describe recursive processes of growth. L-systems have been widely used to draw computer graphics images of many types of plants of both real and imaginary species. L-systems and extended frameworks are very useful not only for drawing but also for understanding formal aspects of morphology by clarifying how wide a variety of shapes a simple rule set can generate.

In real biological organisms, the rule set for developmental processes is encoded on the chromosomes carrying the genotype, which have changed from simple to sophisticated form through several billion years of evolutionary process while adapting to the environment. It will be helpful to combine models of evolution

---

1991 *Mathematics Subject Classification*. Primary 92-08, 92D15; Secondary 92B99, 92D40.

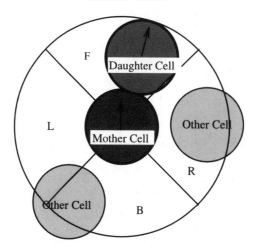

FIGURE 1. Spawning a daughter cell in 2D space. Environment:
$FLBR = 0011_2 = 3$.

and development from a standing point of *Artificial Life* [**LA**] in which we approach the intrinsics of life through synthesis. We designed a model of evolution of development processes, and examined it with computer simulation.

For real natural organisms, various types of features of physical and chemical entities and events affect the cell activities. To avoid the complicated tasks of building a detailed realistic model, many types of physical features are ignored such as gravity, shade, weather, seasons, and so on. All of our models presented below do not use discrete grid worlds but continuous two-dimensional (2D) and three-dimensional (3D) Euclidean space because it theoretically provides infinite freedom to form a shape.

The following sections describe our proposed models, simulation and results, and some remarks.

2. Morphology

Speaking at an abstract level on the developmental process, each cell decides its action according to the rules encoded in its genome, conditioned by its own status and local environment. For mathematical modeling of cell division, we assume that the orientation of division is determined by these two kinds of status, the inner state and the local environmental state, and genetic information the cell contains. Because of the difficulty of simulating all of these complicated features, we assumed

1. the cell shape is a circle in 2D space or a sphere in 3D space,
2. the cell size is constant,
3. cells do not split but spawn daughter cells at an adjoining side,
4. cells do not move from the original position where they were born,
5. cells may spawn daughter cells only if there is enough empty space, and
6. each cell has its own direction as one of the attributes.

A cell has an attribute indicating whether it is active or inactive. Each active cell 'intends' to spawn its daughter cell at an adjoining side where a 'gene' corresponding to the current state designates the relative orientation. In the 2D version, the environmental state is represented by a four-bit integer as shown in Figure 1 in

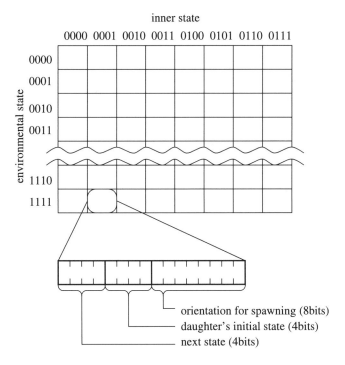

FIGURE 2. Form of chromosome and gene for 2D version.

which the circular area around the cell is divided into four fan-shaped areas: *front*, *back*, *left* and *right*. A value 1 bit indicates that the center of at least one other cell exists in the corresponding fan-shaped area.

The inner states are also represented by four-bit integers, where the most significant bit indicates whether the cell is active (0) or inactive (1).

The conditional part of the development rules consists of a pair of instantaneous values of inner and environmental states. The action part contains relative orientation from the cell's direction to spawn a daughter cell, the daughter's initial inner state, and next inner state of the cell itself. Actually, we employ a look-up table to represent these rules, that is an array of actions where the row indicates environmental state and the column indicates inner active state as shown in Figure 2. The relative angle information is encoded in an eight-bit integer proportionally translated into a real value from $-\pi$ to π. Thus, each table entry includes $4 + 4 + 8 = 16$ bits, the table size is $8 \times 16 = 128$, and so in total one genome consists of $16 \times 128 = 2048$ bits.

To extend this into 3D space, for each look-up table entry, to decide the orientation of the daughter cell we used a triplet of angles as shown in Figure 3. Thus, the action part of each rule includes more two eight-bit integers, that is totally $4 + 4 + 8 \times 3 = 32$ bits. Our current model for the 3D version does not use environmental state but only inner state to reduce the memory size. Thus, one genome consists of $32 \times 8 = 256$ bits. We should consider six more bits of information for environmental state in 3D space, adding *upper* and *lower* to the four directions in 2D space, which would increase the table size by a factor of $2^6 = 64$.

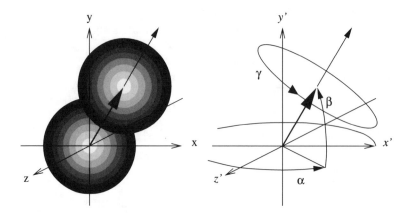

FIGURE 3. Spawning a daughter cell in 3D space. α and β indicate the position on the surface of mother cell, and γ is the argument of the daughter's direction around the mother-daughter axis. A daughter always adjoins tangentially to the mother cell, so this axis is along the line joining the centers of the two cells.

All active cells 'inteond' to spawn daughter cells simultaneously. There may occur some conflict when more than two cells are going to occupy common empty space for their daughters. In such a case, our simulator gives higher priority to a younger cell.

3. Evolution

The evolutionary process for natural organisms strongly depends on the inter-species interactions such as the food web. Starting from preliminary settings without any ecological features, we examined evolution by a Genetic Algorithm (GA) [G] with pre-defined fitness function. In the 2D version, we make the fitness proportional to the number of cells after a fixed number of steps of growth, however the fitness value is forced to be zero if the individual is still able to grow after the final step of the generation. The reason why we added the latter criterion is that unrestricted growth brings explosive growth without forming any interesting shapes. In the 3D version, the fitness f is defined as

$$(3.1) \qquad\qquad f = \frac{n}{0.1d^2 + 0.2d + 0.7\sqrt{d}}$$

where n is the number of cells and d is the distance between farthest cells. These definitions were designed so that there emerge some interesting shapes such as spirals and other types of fractal forms. The reason why the fitness measure of 2D was not employed in the 3D case is that it is harder to stop growth by obstruction between cells: The 3D case needs a plane to prevent another branch from growing, while the 2D case requires only a curved whisker. According to our preliminary experiments, the fitness measure of 2D always leads to explosive growth in 3D space.

We use a type of generational GA in which every individual in the population is initialized with a random genotype and is tested through a selection process to decide whether it remains in the next generation or not. To accelerate the

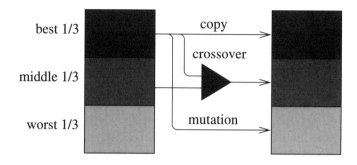

FIGURE 4. 1/3 selection.

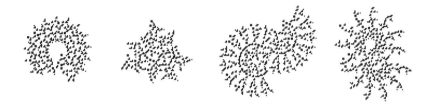

FIGURE 5. Examples of the typical shapes in 2D version of the
best individual on distinct evolutionary processes using different
random number sequences.

evolutionary process, we did not employ ordinary selection algorithms widely used
in GA, such as the roulette-wheel selection, ranking selection, nor other probability-
based selection mechanisms, but rather the *1/3 selection* algorithm as shown in
Figure 4. The details are as follows.

1. The best third of the population remains in the next generation without any
   modification of genotype;
2. the middle third of the population is replaced with individuals generated
   using crossover operation between the best third individuals and the middle
   third individuals; and
3. the worst third of the population is replaced with mutants of the best third
   individuals.

Figure 5 shows some typical phenotypic shapes of the best evolved individuals
in 2D space with forty steps in one generation. Usually we can find shell-like
shapes after some generations pass, because these types represent a good strategy
for stopping growth on time while maximizing the number of cells simultaneously.

Figure 6 shows two typical shapes of best evolved individuals in 3D space. We
can find stick-like shapes in an early stage of the evolution since it is not necessary to
stop growth as described above, but they always disappear soon when ball, spiral,
plate-like shapes appear because of d in the denominator in equation 3.1. This
definition of fitness avoids stick shapes, and spiral and plate do comparably to ball
shape when d is small.[1]

---

[1] Cell diameter is 0.2 in our experimental simulation.

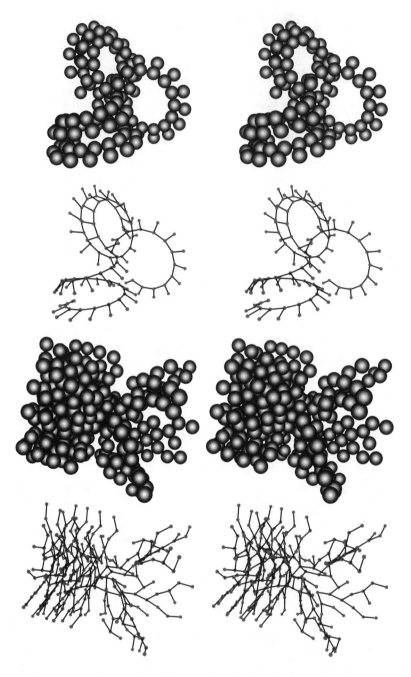

FIGURE 6. Examples of best evolved individuals in the 3D version (stereograms). The lower images of each pair are skeletons on which the cells appear. Each segment of the skeletons connects between the center points of a mother cell and its daughter cell.

4. Ecology

The above assumption of no interaction among individuals and a pre-defined fitness measure is unsuitable for a model of the evolutionary process of natural organisms, because natural fitness depends on how many descendants each individual spawns and can be known only after some generations as the result of the evolutionary process.

We examined introducing an ecological feature, that is, interaction among individuals. All of the individuals in the population grow in a common space from randomly placed seeds. We assumed one inactive state, 1111_2, indicates a *seed* that only becomes active in the next generation. After a fixed number of growth steps, all of cells except seeds die away and the growth process restarts again from the remaining seeds, whose state is forced to be 0000_2, however the total number of seeds of one individual is restricted by the size of its grown body. This setting is just like a life cycle of grasses which sprout in spring, grow in summer, wither in fall, and pass the winter as seeds.

Figure 7 shows an example of distribution of grown organisms at the fifth generation where each generation includes fifty steps. In this case, there are 294 individuals and totally 13,287 cells. An individual without plural seeds cannot propagate even if it grows into a large body. A large body that can distribute the seeds over a wider area achieves successful propagation, because it wins in the competition for rapid space occupation among the individuals. It is usually observed that natural plants that have survived through the course of evolution have the ability to widely distribute offspring by spreading branches, and employing wind, birds, and insects, etc., to carry their seeds.

5. Conclusion

The results of simulation presented above are only some samples of figures we found. Though it provides one possible kind of evidence concerning the diversity of forms that have emerged through evolution, we should do more investigation on the process of evolution and the effects of some parameters to draw out results fruitful from the viewpoint of biologists.

Directions to extend the model described above are also considerable, such as including simulation of physics, metabolism and ecology. Combination with other morphological research methods such as those of [**F**] and [**O**], and artificial botany approaches such as those of [**C**] might provide us good candidates to progress in this research in the near future.

These types of models can be used, not only as an A-Life (Artificial Life) facility to investigate the evolution of developmental systems, but also can be seen as a form of automatic computer graphics art. The 2D version of the system discussed here was shown at the A-Life Art Exhibition at the Tokyo International Museum in 1993. The approach may have good applications in the field of entertainment.

References

[C] R. L. Colasanti, R. Hunt, Real Botany with Artificial Plants: A Dynamic, Self-Assembling Plant Model for Individuals and Populations. In: P. Husbands, I. Harvey (eds.), *Proc. of the Fourth European Conf. on Artificial Life*, MIT Press, (1997), 266–273.

[F] K. Fleischer, Investigations with a Multicellular Developmental Model. In: C. G. Langton, K. Shimohara (eds.), *Artificial Life V*, MIT Press, (1996), 229–236.

FIGURE 7. Examples of distribution of grown organisms at the fifth generation through ecological selection starting from 80 randomly positioned seeds of randomly generated genotypes. Each generation consists of 50 steps. The size of cell is 20 units and the field is an 800×800 unit torus.

[G] D. E. Goldberg, *Genetic Algorithms in Search, Optimization and Machine Learning.* Addison-Wesley, (1989).

[LA] C. G. Langton, Artificial Life. In: C. G. Langton (ed.), *Artificial Life*, Addison-Wesley, (1989), 1–48.

[LI] A. Lindenmayer and P. Prusinkiewicz, Developmental Models of Multicellular Organisms: A Computer Graphics Perspective. In: C. G. Langton (ed.), *Artificial Life*, Addison-Wesley, (1989), 221–249.

[O] A. Onitsuka, J. Vaario, K. Ueda, Structural Formation by Enhanced Diffusion Limited Aggregation Models. In: C. G. Langton and K. Shimohara (eds.), *Artificial Life V*, MIT Press, (1996), 237–251.

DEPARTMENT OF INFORMATION SYSTEMS SCIENCE, SOKA UNIVERSITY, HACHIOJI, TOKYO 192-8577, JAPAN

E-mail address: unemi@iss.soka.ac.jp

Lectures on Mathematics in the Life Sciences
Volume **26**, 1999

Computer Simulation and Analysis of a Growing Mammalian Cell Colony

V. V. Savchenko, A. G. Basnakian, and A. A. Pasko

ABSTRACT. All living creatures are made of cells - small membrane-bounded compartments filled with a concentrated aqueous solution of chemicals. This paper deals with simulation and characterization of a growing mammalian cell colony. A colony is modelled as a set of contacting deformable particles. Variation of size, orientation and form of a cell is assumed, although ellipsoidal shape of a cell is used as a basic form. Modelled cells were compared to AHH-1 human lymphocyte culture cells by visual inspection and with the help of the fractal analysis methods.

1. Introduction

The problem of computer simulation for creating artificial life has been widely studied lately. Biology is one of the main sources of complex problems that require computer simulation. This paper deals with simulation of a growing mammalian cell colony. A colony is modelled as a set of contacting deformable particles. A particle can be substituted by a pair of new particles, modelling the process of cell division. From the geometric modelling point of view, this application has identified a complex problem of modelling arbitrarily shaped geometric particles, their collision and packing. Before a computer-simulated model of a colony can be built, we have to develop algorithms for simulation of dynamic behaviour of particles and communication between them, controlling particles geometry and rendering.

The main goals of this work are:

- To provide better understanding of dynamical characteristics of the cell reproduction process,
- To apply a geometrical approach to modelling "communications" between particles,
- To study cell growth as a property of behavior rather than as a property of internal physiology,
- To address how can such a simulation be validated.

The remainder of the paper develops as follows. The biological origin of the model is observed in the next section. Section 3 is devoted to the related works.

1991 *Mathematics Subject Classification.* Primary 68U05, 92–04, 65K10, 68U20, 70F35.

The second author was supported by ORISE/NCTR fellowship from the Department of Energy, USA.

Section 4 describes our algorithm for colony growth simulation. Section 5 evaluates biological adequateness of the model.

2. Biological origin of the model

All living creatures are made of cells - small membrane-bounded compartments filled with a concentrated aqueous solution of chemicals. In all cells of higher plants and animals, mammalians and human, the bulk of hereditary material - DNA - is isolated in a membrane-bound nucleus lying in cytoplasm [**Tr**]. The nucleus is regularly ovoid or spherical and darkly staining, while the surrounding cytoplasm is more lightly staining [**Fre**]. The high density, size and shape of the nucleus are much more conservative than those of cytoplasm because the nucleus contains highly condensed genetic material. The nucleus usually is in the range of 3 to 14 microns in diameter, and most of mammalian cells are in the range of 5 to 50 microns. There is great variation of the appearance, size, and shape of cells, which reflects the different functions of different cell types. There is also dependency on the location of a cell and its neighborhood. The nucleus is usually located close to the cell center. Both the nucleus and the entire cell are enveloped by a membrane that is only 7.5 nm thick, and thus is often invisible under the light microscope.

Isolated from the organism, cells can be grown *in vitro* on a plastic or glass surface in the presence of liquid media containing all necessary ingredients [**Fre, Ha**]. These are so-called cell cultures. In cell culture, cells are growing exponentially by cell division (duplication) while all necessary nutritional components are available, so that a single cell plated onto plastic surface after a certain number of generations forms a colony of cells. One colony consists of several cells to several thousand cells. Single cell division involves division of the cytoplasm which follows immediately after the division of the nucleus [**Al**]. In most cells, the division of the nucleus occurs by mitosis to ensure that each of two daughter cells gets equal DNA genetic content identical to that of the parent cell. Normal (non-tumour) cells usually form flat single-level colonies and have a limit of about 40 generations *in vitro* (Hayflick's limit) [**Hay**]. Tumour cells have no such limit, and are able to grow without obligatory single-layer growth and form multi-level colonies. Although cells in a colony can perform slow amoebic movements, cell-to-cell adhesion (cell junctions) is quite resistant to mechanical forces tending to separate cells. It was indicated that spaces between cells or between a cell and a surface are narrow, on the order of 15 to 25 nm. Currently there are several thousand cell lines used in research and industry [**Ha**], but the biology of cell colonies, namely the dynamics of their growth, is still not enough investigated.

3. Related works

In the following overview we mention related work of several types: artificial life, non-rigid bodies modelling, collision detection and packing. Artificial life endeavors to synthesize lifelike behaviours within computers. The key parameters for creating artificial life in virtual worlds have been identified by the authors of [**Th**]. They notice that very simple creatures have been discussed and there has been little visual representation of living organisms in previous work. The authors of [**Ra**] mentioned that simulation techniques of cell division processes have their origins in the cellular automata. The authors use graph theory to represent cell division. The cellular automata approach is used to simulate botanical colony growth in [**Ku**].

Now, various approaches should be integrated to create truly virtual worlds with autonomous living beings, plants, animals and humans with their own behaviour, and real people should be able to enter into these worlds. Paper [**Fra**] presents methods of subdivisions or partitions of geometric spaces into cells (vertices, edges, faces, etc.) and basic operations to define sub-object behaviour.

The survey [**Ga94**] presents the state of the art in animation of non-rigid bodies. In particular, models based on the physical theory of elasticity in continuous media and discrete mechanical models which integrate discrete mechanical components are observed. This survey demonstrates some optimization techniques to specify constraints on the behaviour of the objects and collision detection and response algorithms. Constructing a virtual world requires the ability to simulate geometric collisions in order to provide collision avoidance, for example, in behavioural procedural modelling [**Re**]. In [**Si**] a particle animation system using data parallel computation on the Connection Machine CM-2 is presented. The author emphasizes that particle systems are well suited for highly parallel computation. Paper [**Lu**] investigated random disk packing of spheres which enlarge in size and offered an effective algorithm for determination of pair collisions. Detecting and eliminating collisions between plant organs while optimizing their packing are presented in [**Fo**], whose proposed method arranges plant structures of varying sizes on arbitrary surfaces of revolution. Motion of time-dependent parametric surfaces is simulated in [**Vo**], where the authors state that the collision detection problem is insoluble for arbitrary time-dependent surfaces. To fill the gap in the collision detection problem for simulation of dynamic behaviour of rigid solids with implicitly defined surfaces, a numerical algorithm for collision detection was used in [**Sa**].

In [**Fl**] an approach was proposed for generating patterns of geometric elements using a biologically-motivated cellular simulation. The cellular development system forms the basis of this work and combines cell-cell interactions, cell-cell adhesion, oriented particles, and surface constraints. The developed system is a powerful method of creating attractive computer graphics models of organic objects. The authors of [**Fl**] noticed that simulations can be slow for some kinds of cell programs, even for the spherical shapes.

Since artificial objects can seem very similar to natural ones, the question, *How can such an object or model be validated?*, appears with some risk to accept similarity as validation. Simulation leaves open the question of how to validate the results and we suppose that simulation has to reproduce experimental data, regularities and features of phenomena. Paper [**Gr89**] presents a good example how it is possible to validate rules and behaviour of complex structure via the help of computing simulation and visualization.

4. Simulation of colony growth

4.1. Particles geometry and intersection. We model living cells by geometric objects called particles. To define a geometric object the function representation is used: both simple and complex objects are represented by inequalities $f(x, y, z) \geq 0$, where f is a real continuous function of Cartesian coordinates of a point. The function has negative values for points outside an object, positive values for inside points and zero values for points on the surface. Surfaces defined by equality $f(x, y, z) = 0$ are usually called implicit surfaces. This is very convenient for defining simple surfaces like spheres, ellipsoids, tori and others. For description of more complex form a set of orthogonal functions, for example, can be used to express a defining function as a set of spherical harmonics. Although the numerical algorithms applied here were designed for arbitrarily shaped objects, in our application we limit ourselves by using a "noisy ellipsoid" as a particle:

$$(4.1) \qquad f(x,y,z) = 1 - (\frac{x}{a})^2 - (\frac{y}{b})^2 - (\frac{z}{c})^2 + noise(x,y,z),$$

where $noise(x,y,z)$ is a continuous stochastic function usually called "solid noise" and represented here by Gardner's series [**Le**].

Points belonging to the interpenetration area of two particles can be detected with the help of the interference relation. This relation is defined by the bivalued predicate:

$$(4.2) \qquad S(f_1, f_2) = \begin{cases} 0 & \text{if } f_1(p)\&f_2(p) < 0 \\ 1 & \text{if } f_1(p)\&f_2(p) \geq 0 \end{cases}$$

where f_1 and f_2 are defining functions of two particles; p is a point with coordinates (x, y, z) in an admissible domain P (a rectangular domain such that all points of intersection belong to it); & is a symbol of set-theoretic intersection. Analytical definitions of set-theoretic operations in the form of so-called R-functions have been studied by [**Rv**]; see also [**Sh**] for a survey. The simplest type of R-intersection is well-known: $f_1\&f_2 = min(f_1, f_2)$.

The following numerical algorithm for detection of interpenetration area is proposed:

1. Searching for the admissible domain P is performed through the intersection test of a couple of bounding boxes according to the following scheme: bounding boxes are projected onto three coordinate planes; projection intersections are determined in each plane; a rectangular domain such that all points of intersection belong to it is found. If this admissible domain is empty we have the case when the search for a collision point is of no use.

2. To find intersection points we use Sobol's quasi-random (or $LP\tau$) points [**So**]. $LP\tau$ points fill n-dimensional space more uniformly than traditional random number generators. Their use can lead at most to an $N^{-2/3}$ error term (where N is a number of trials) in Monte Carlo integration applied to estimate an interpenetration area and the position of the center of mass of a particle.

A particle deformed by contact with its neighbors can be described as:

$$(4.3) \qquad \tilde{f}_1(p) = f_1(p) - \sum_{j=2}^{k} f_j(p)$$

where f_1 is the defining function of an initial particle, $\tilde{f}_1$ is the defining function of the deformed particle and f_j $(j = 2, \ldots, k)$ are defining functions of neighbor particles. Such a description of deformation was proposed in [**Ga93**]. It has no physical meaning but it is enough to simulate deformations when contacting objects are described by functions of the same type.

4.2. Computer model of a growing cell colony. Mechanical modelling of a growing mammalian cell colony is supposed to provide a better understanding of dynamical characteristics of the cell reproduction process. We believe that it is possible to study cell life as a property of behaviour rather than a property of internal physiology. Also, we realize that there is a huge number of factors which may directly cause or just influence cell division and colony growth. Some of these factors have been chosen as important ones for colony growth. The exact numerical data used in the model were based on experimental data obtained from cell biology literature for the "average" human or mammalian cell growing in a culture *in vitro* on some kind of surface. In order to study the opportunity of applying a geometrical approach to formation of structure, and to estimate time performance of the numerical algorithm, we consider a simplified mechanical model of human cell colony growth. In the model, we take into account contact guidance and inhibition to control a preferred direction of cell motion as major factors. We consider that a living cell can be in two states: active (dividing) or passive (non-dividing). At the stages of colony growth which we take, cell death does not occur.

We use the following rules of cell reproduction:

- reproduction of cells is a result of cell division;
- during a single division, a parent cell forms a daughter cell, i.e. a particle is substituted by a pair of new particles;
- the initial number of cells in a colony is one, maximal number is not defined;
- the number of feasible divisions of the initial cell is limited by 40 (Hayflick's limit, see above);
- for cells with open boundaries division occurs every 12 ± 2 hours;
- division does not occur if there is not enough space for both parent and daughter cells;
- cell division is presented by splitting both nucleus and cytoplasm by a plane passing through the nucleus center and a point with minimal distance between the nucleus and cytoplasm membrane; orientation of the splitting plane is found by a genetic free space search algorithm (see below); a straight line orthogonal to the splitting plane defines the principal splitting direction, which can be modified by the genetic search;
- after division, the cytoplasm is split in two unequal parts and a nucleus in a daughter cell reaches the size of the one of a parental cell;
- within two hours after division a nucleus occupies a position with maximal distance from the cell's faces;
- variation of size, orientation and form of a cell is assumed, although ellipsoidal shape of a cell is used as a basic form;
- "solid noise" is used for attaching irregularity to a cell's shape;
- representation of a cell as a deformable (soft) particle assumes formation of faces between intersecting particles;
- cells that do not intersect with neighbors have random splitting direction.

In the outside physical world there are a lot of initially simple geometric forms that can produce complex forms in accordance with certain rules or principles. In our model, we use the "economy" principle: dividing cells strive to occupy space with the minimal deformation of each other. This principle defines the measure of fitness in the free space search. Formulation of the numerical simulation algorithm follows.

Let the 3D domain, initial position and geometry of a cell which can be divided be given. We want to study formation of the shape of a cell colony as a function of time. Global state $S(t)$ of the system at time t is the vector: $(s_1(t), s_2(t), ..., s_N(t))$, where N is a number of cells in a colony, $s_i(t) =$

$(r_i, r\_c_i, A_i, B_i, C_i, W_{i1}, W_{i2}, W_{i3}, T_i)$ is the state of the i-th cell; r_i is the center position vector of the cell; r_c_i is the center position vector of the nucleus; A_i, B_i, C_i are semiaxes of the ellipsoid; W_{i1}, W_{i2}, W_{i3} are angular velocities of rotation for computing orientation of the cell; and T_i is time of last splitting of the given cell. We use the discrete-event approach to advance the state of the system. The discrete-event approach means changing of the state in particular time moments when an event (division) takes place. We use a global time step as a minimal time interval for advancement from event to event. At each time step we define free space, new positions and shapes of parent and daughter cells and calculate positions of nuclei that define splitting direction at the next splitting step. A nucleus is moving in a cell after splitting for two hours. Since this time is less than splitting time, we recalculate the position of each nucleus only once after splitting. We calculate the position of a nucleus and possible direction of splitting according to the following scheme:

- Modelling interpenetration area. As it was mentioned above, if particles i and j interact, we can define interpenetration area by $f_{ij} = f_i \& f_j$. Volume V_{ij} of this area is estimated with Monte Carlo integration. It is used to evaluate a fitness function in the free space search and packing.
- Computing a center of mass. In order to calculate new position of a nucleus we use Monte Carlo integration. In this case volume of a particle deformed by its neighbours is calculated.
- Search for a point with minimal distance between new center of mass and the cell's cytoplasmic membrane. We select an $LP\tau$ point which does not belong to the deformed particle and has minimal distance from the center of mass. A plane passing through this point and the center of mass defines the splitting direction.

If a cell can reproduce, it has to define space for a daughter cell. This search for free space looks like communication with the cell's neighborhood. A cell is trying to find free space by pushing all neighboring cells. Since a colony is continuous, only deformations of cells are allowed in our model. We simulate deformation by interpenetration of cells with a preassigned measure of tolerance. To search for free space, new positions and shapes of parent and daughter cells, we apply a genetic algorithm (GA). GAs can be attractive in applications work for several reasons [**Go**]:

- GAs can solve sophisticated problems quickly and reliably;
- GAs are easy to interface to models;
- GAs are extensible.

There are many variations of genetic algorithms. A GA is a mathematical search technique based on the principles of natural selection and genetic recombination. The simple genetic algorithm developed by [**Ho**] begins with a set of random structures as possible solutions to the given problem. The structures are evaluated using a fitness weighted random selection scheme, and by applying genetic operators such as mutation and crossover to them. The resulting structures are then evaluated, the new structures with higher fitness join the population to replace those old ones whose fitness measures are lower. Therefore, the structures with high strength tend to survive and those with low strength to die out. The process is repeated until a satisfying structure (an optimal solution to the given problem) is obtained. In the genetic algorithm, binary strings represent possible candidate solutions. Each

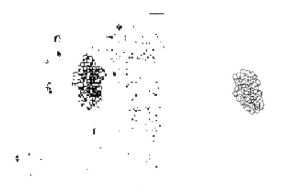

FIGURE 1. Images of AHH-1 human lymphocyte culture cells (left) and modelled cells (right) after contours extraction.

string position takes a value of 1 or 0. In this context, a string is composed of information encoding for modelling parameters $r_i, A_i, B_i, C_i, W_{i1}, W_{i2}, W_{i3}$. These parameters for a new cell are represented by a bit string, from which they can be decoded into real values. A fitness function in GA represents the environment for all individual strings. It provides measurement of how good a given solution is. Usually, the probability of reproduction during a given cycle is proportional to the fitness of the individual string. The effect of this is that individuals with higher fitness have a higher expected number of offspring than individuals with lower fitness and have a greater chance to survive. The fitness function in our simulation has the following form:

$$(4.4) \qquad Fitness = (V_i - \sum_{j=1}^{k} V_{ij})/V_i$$

where V_i is volume of a new cell, V_{ij} is volume of interpenetration area of the new cell with j-th neighbouring cell. The fitness function has a non-negative real number value in the range of [0,1]. For a new cell to be generated at time t, the GA is used to search for an optimal solution which has highest fitness at that time.

Conceptually the algorithm consists of two steps: 1) a predictor which provides computation of a daughter cell and its fitness; 2) a corrector (works only if predictor was unable to generate a daughter cell with "good" fitness) which provides new generation calculations applying slight derivations of the cell obtained from the predictor, new fitness computation for each generation, best cell selection between those generations, new center of a nucleus computation for the selected cell. Parallelization of the predictor step is a way to increase performance of the algorithm [Sa95].

5. Evaluation of biological adequateness of the model and characterization of surface topography

Modelled cells (Fig. 1, right) were compared to AHH-1 human lymphocyte culture cells (Fig. 1, left). Visually, modelled cells are a little different from real cells. They are not transparent, darkly coloured, and have an intensive dark shadow. Under the microscope, cells are transparent and have no shadow, because they are illuminated from the bottom not from the side. From the cell biologist's point of

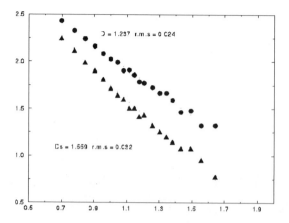

FIGURE 2. Box counting results for the images shown in Fig. 1.
The plot of log N(r) vs. log r is linear

view, the model has several definite advantages. Cell shape and size, form of colony
and its dynamics are very similar to those of real cells. Modelled cells are different in
their size, expressing the variability of forms which is a common feature of biological
objects [**Fre**]. A cell is spread on the surface, and neither its border nor cover are
even. It is visible, that in the center of colony cells are smaller than those on the
colony edge. Central cells seem to be restricted in space. It is important, that
cells which are standing separately from the visual colony border are significantly
different in their sizes and shapes. Cells in a colony are able to perform amoebic
movements, which lets them stand almost separately from the colony, especially in
young colonies. These movements cause some very realistic open spaces in young
colonies. Also, cells in the division process, form characteristic doubling cells with
a straight border between them. Colonies do not form secondary layers of cells,
but instead the rate of cell division goes from the exponential (logarithmic) phase
to contact inhibition and density limitation of growth, very characteristic for non-
transformed (non-tumour, normal) cells [**Ha, Tr**]. As in a real colony, there is
always a mixture of dividing and resting cells. Even dividing cells are different
in their sizes. The average cell size in an exponentially growing population is
decreasing. The size of a colony as well as the number of cells in it is of course
growing. But, what is important, the colony is changing its shape while growing,
and is not at all round. The colony itself is flat (normal cells) but has non-even
border. All these characteristics are often observed in the literature [**Fre, Tr, Ja**]
and in our experiments with AHH-1 cells.

   We clearly understand that human vision or estimation can not be sufficient
or complete reason to draw conclusions about model correctness. The main question about the biological adequateness of the model, *Can a relevant statement be
made by visual inspection?*, is left open. The same question arises about the use
of computers. Digitized images represent foreground and background pixels with
some noise, brightness and color. They require careful correction for preserving the
details of a studied phenomenon. Nevertheless, we suppose that some measurement

parameters, for instance, the fractal dimension, correlate with the properties of the real-world phenomena. Fractal dimension also seems correlated with human visual response to simulated properties. We use the fractal dimension as a characteristic of objects in a general manner. Different types of real or simulated cells can have significantly different fractal dimensions. We suppose that such differences may allow us to detect the correlation between the real colony and simulated one.

There are a number of numerical methods that can be applied in practice to calculate the fractal dimension and to indicate its existence [**Vi, Pe**]. The box counting method and the information dimension method were implemented. Calculated results (Fig. 2) of the box counting method can be approximated by a straight line with error limits from 0.02 to 0.03. In the box counting method, the number of non-empty squares $N(r)$ is calculated, where r defines the image spacing. The second method gives slight different but qualitatively similar results.

We would stress that the analysis shows fractal scaling and exhibits some hidden features so often found in the real world. To make a reliable conclusion, in future we have to achieve visual similarity in the model to real cells, making them transparent with visible nuclei. Appropriate image processing of experimental and simulated data should be carefully done to avoid bias in the density histograms.

References

[Al] B. Alberts, D. Bray, J. Lewis, M. Raff, K. Roberts, and J.D. Watson, *Molecular Biology of the Cell*, Garland Publishing Inc., (1983).

[Fl] K.W. Fleischer, D.H. Laidlaw, B.L. Currin, A.H. Barr, Cellular texture generation, *SIGGRAPH 95*, Los Angeles, California, August 6-11, (1995), 239–248

[Fo] D.R. Fowler, P. Prusinkiewicz, J. Battjes, A collision-based model of spiral phyllotaxis, *Computer Graphics*, vol. 26, No. 2, (1992), 361–368.

[Fra] J. Francon and P. Lienhardt, Basic principles of topology-based methods for simulating metamorphoses of natural objects, *Artificial Life and Virtual Reality*, N.M. Thalmann and D. Thalmann (Eds.), John Wiley & Sons Ltd, (1994), 23–44.

[Fre] R.I. Freshney, *Culture of Animal Cells*, Wiley-Liss, (1987).

[Ga93] M.-P. Gascuel, An implicit formulation for precise contact modeling between flexible solids, *Computer Graphics Proceedings*, Annual Conference Series, (1993), 313–320.

[Ga94] M.-P. Gascuel, C. Puech, Dynamic animation of deformable bodies, *From Object Modelling to Advanced Visual Communication*, S. Coguillart *et al.* (Eds.), Springer-Verlag, (1994), 118–139.

[Go] D.E. Goldberg, Genetic and evolutionary algorithms come of age, *Communications of the ACM*, vol. 37, No. 3, (1994) 113–119.

[Gr89] N. Green, Voxel Space Automata: modelling with stochastic growth processes in voxel space, *Computer Graphics*, vol. 23, No. 3, (1989), 175–184.

[Ha] R. Hay, J. Caputo, T.R. Chen, M. Macy, P. McClintock, and Y. Reid (Eds.) *American Type Culture Collection Catalogue of Cell Lines and Hybridomas*, Seventh edition, ATCC Publishing, (1992).

[Hay] L. Hayflick and P.S. Moorhead, The serial cultivation of human diploid cell strains, *Exp. Cell Res.*, vol. 25, (1961), 587–621.

[Ho] J.H. Holland, K.J. Holyoak, R.E. Nisbett and P.R. Thagard *Induction: Processes of Inference Learning and Discovery*, MIT Press, (1986).

[Ja] S.J. James, A.G. Basnakian, and B. Miller, *In vitro* folate deficiency induces deoxynucleotide pool imbalance, apoptosis, and mutagenesis in Chinese hamster ovary cells, *Cancer Res.*, 54, (1994), 5075–5080.

[Ku] T.L. Kunii, Y. Takai, Cellular self-reproducing automata as a parallel processing model for botanical colony growth pattern simulation, *New Advances in Computer Graphics, Proceedings of CG International'89*, R.A. Earnshaw and B.Wyvill (Eds.), Springer-Verlag, (1989), 7–22.

[Le] J.P. Lewis, Algorithms of solid noise synthesis, *Computer Graphics*, vol. 23, No. 3, (1989), 263–270.

[Lu] B.D. Lubachevsky and F.H. Stillinger Geometric properties of random disk packing, *Journal of Statistical Physics*, vol. 60, Nos. 5/6, (1990), 561–583.

[Pe] H. Peitgen, H. Jurgens, D. Saupe, *Chaos and Fractals*, New Frontiers of Science, Springer-Verlag, (1992), 984 p.

[Ra] R. Ransom and R.J. Matela, *Computer Graphics in Biology*, Croom Helm Ltd, London and Sidney, (1986).

[Re] C.W. Reynolds, Computer animation with scripts and actors, *Computer Graphics*, vol. 16, No. 3, (1982), 289–296.

[Rv] V.L. Rvachev, *Methods of Logic Algebra in Mathematical Physics*, Naukova Dumka Publisher, Kiev, (1974). (in Russian)

[Sh] V. Shapiro, Real functions for representation of rigid solids, *Computer Aided Geometric Design*, vol. 11, No. 2, (1994), 153–175.

[Sa] V. Savchenko, A. Pasko, Simulation of dynamic interaction between rigid bodies with time-dependent implicitly defined surfaces, *Parallel Computing and Transputers, Proc. PCAT-93*, D. Arnold *et al.* (Eds.), IOS Press, (1994), 122–129.

[Sa95] V. Savchenko, A. Basnakian, A. Pasko, S. Ten, R. Huang, Simulation of a growing mammalian cell colony: collision-based packing algorithm for deformable particles, *Computer Graphics: Developments in Virtual Environments*, R. Earnshaw and J. Vince (Eds.), Computer Graphics International'95 Conference, Academic Press, (1995), 437–447.

[Si] K. Sims, Particle animation and rendering using data parallel computation, *Computer Graphics*, vol. 24, No. 4, (1990), 405–412.

[So] L.M. Sobol, *Numerical Monte Carlo Methods*, Nauka Publisher, Moscow, (1986). (in Russian)

[Th] N.M. Thalmann and D. Thalmann, Introduction: Creating artificial life in virtual reality, *Artificial Life and Virtual Reality*, N.M. Thalmann and D.Thalmann (Eds.), John Wiley & Sons Ltd, (1994), 1–10.

[Tr] J.P. Trinkaus, *Cells into Organs*, Prentice-Hall Inc., (1984).

[Vi] T. Vicsek, *Fractal Growth Phenomena*, 2nd ed. World Scientific, (1992), 488 p.

[Vo] B. Von Hertzen, A. Barr, H. Zatz, Geometric collisions for time-dependent parametric surfaces, *Computer Graphics*, vol. 24, No. 4, (1990), 39–48.

VLADIMIR V. SAVCHENKO, PROFESSOR, LABORATORY OF SHAPE MODELING, DEPARTMENT OF COMPUTER SOFTWARE, THE UNIVERSITY OF AIZU, AIZU-WAKAMATSU, FUKUSHIMA, 965-80 JAPAN
E-mail address: savchen@u-aizu.ac.jp

ALEXEI G. BASNAKIAN, RESEARCH ASSOCIATE, UNIVERSITY OF ARKANSAS FOR MEDICAL SCIENCES, DEPARTMENT OF INTERNAL MEDICINE, LITTLE ROCK, AR 72205, USA
E-mail address: Abasnakian@aol.com

ALEXANDER A. PASKO, ASSISTANT PROFESSOR, LABORATORY OF SHAPE MODELING, DEPARTMENT OF COMPUTER SOFTWARE, THE UNIVERSITY OF AIZU, AIZU-WAKAMATSU, FUKUSHIMA, 965-80 JAPAN
E-mail address: pasko@u-aizu.ac.jp

Lectures on Mathematics in the Life Sciences
Volume **26**, 1999

Why Do Asexual and Self-Fertilising Populations Tend to Occur in Marginal Environments?

Joel R. Peck, Jonathan M. Yearsley and David Waxman

ABSTRACT. We present a mathematical model of a heterogeneous environment in which a sexual population and an asexual population compete. We assume that resources are most available at the centre of the environment, so that the central area can support a higher population density than can be supported on the margins. We find that, if asexuals have an intrinsic fertility advantage, then they tend to take over the entire environment, and the sexual population goes extinct. If sexuals have an intrinsic fertility advantage, then, at equilibrium, they are usually present in the environment, but the asexuals are also often present. When both sexual and asexual populations are present at equilibrium, the sexuals usually have a more narrow range than the asexuals, and the sexuals also tend to occur closer to the resource-rich centre of the environment. Similar patterns are found in many natural environments.

1. Introduction

Many plants and animals reproduce asexually. Asexuality occurs within a wide variety of taxa, including vertebrates such as fish and reptiles (Bell 1982; Maynard Smith 1978). Most multicellular asexual organisms are closely related to species in which some (or all) offspring are produced by sexual means. It has long been known that, if one compares the geographic distributions of closely related sexual and asexual organisms, certain patterns are apparent. These patterns are called geographic parthenogenesis (Vandel 1928). One of the observations commonly reported is that asexual organisms tend to occur in resource-poor environments, which are often on the margins of the ranges of closely related sexual organisms. This pattern has been reported in both plants (Bierzychudek 1985) and animals (Bell 1982; Glesener and Tilman 1978; Hughes 1989; Lynch 1984; Parker 1998; Suomalainen 1950; Vandel 1928). We will call this pattern *asex outside* for the sake of brevity.

Here, we present the results of a theoretical study which suggests that basic ecological and evolutionary processes will tend to generate the pattern of asex outside whenever sexual and asexual species coexist, so long as selection pressures change from place to place. The results also suggest that an asexual species will tend to have a larger geographic range than a similar sexual species, and this pattern is in accord with much of the data (Bierzychudek 1985; Lynch 1984; Parker 1998; Suomalainen 1950). (Note: for convenience, we will refer to asexual "species" despite the absence of mating in these organisms). As we have shown elsewhere (Peck *et al.* 1998), processes similar

1991 *Mathematics Subject Classification.* Primary 92D25.

to the one we describe here may account for other geographic patterns, such as the tendency of asexuals to occur at high altitudes and far from the equator (Bell 1982; Bierzychudek 1985; Glesener and Tilman 1978; Hughes 1989; Suomalainen 1950; Vandel 1928).

In many groups of organisms, a variety of types of reproduction are found. These may include outcrossing (mating with non-relatives), selfing (the extreme of inbreeding, where males and female gametes come from the same individual) and asexuality (where each individual has only one parent and there is no recombination, so that parents are identical to their offspring, except for mutations). It appears that highly-selfed organisms tend to have geographic distributions that are similar to the distributions of asexuals. That is, selfers tend to occur in more marginal and resource-poor environments than closely related outcrossing organisms (Barrett 1988; Jarne and Charlesworth 1993; Lloyd 1980). As we shall see, our model of asexuality can also be used to represent a highly-selfed organism. Thus, the results presented here can help to explain the geographical distributions of outcrossers and selfers.

Jannis Antonovics has shown that, when one small part of a habitat has a very different environment from the rest of the habitat, self-fertilisation can be favoured in this small and environmentally different area because selfing can lead to protection from mating with locally maladapted migrants and their descendants (Antonovics 1968). Here, we show that a similar process can lead to selfing and asexuality in marginal areas, even in the more-general situation where the environment changes in a gradual way from place to place. This is because the relatively small populations in marginal areas are subject to extraordinarily high rates of migration from the more densely populated centre of the environment.

2. Methods

To construct our model, we assume the existence of two ecologically similar species, one a sexual hermaphrodite, and the other asexual. We assume further that both species are present in sufficiently large numbers so that stochastic processes can be ignored (except when a species is on the verge of extinction). Both species are annuals, so that the adults die shortly after the birth of their offspring.

We consider a long and narrow habitat which stretches from east to west. Position within this habitat is denoted by x. At the centre of the habitat, we set $x=0$. As we move to the east of the centre, the value of x increases to indicate our new position, taking on values of $x=1$, $x=2$, $x=3$, up to $x=\Omega$, which is the most easterly habitable position in the environment. As we move to the west from the centre, x takes on values of $x=-1$, $x=-2$, ... $x=-\Omega$. We carried out the analysis of our model using computer simulations. Therefore, it was convenient to assume that x takes on only integer values. This means, essentially, that the environment consists of discrete patches arranged in a line from west to east.

In order to produce offspring, adults must consume resources. Let $R(x)$ represent the density of resources at a particular location, x. Thus, in the case of a plant, $R(x)$ might represent the density of free nitrogen in the soil, and in the case of an animal, $R(x)$ might represent the density of edible plants. Let $R_0(x)$ represent the value of $R(x)$ at the beginning of a generation. For simplicity, let us assume that, for each value of x, the value $R_0(x)$ is the same at the beginning of each generation. This suggests that resources are renewed each generation, and that any resources remaining at the end of a generation are degraded, so that they do not accumulate.

We assume that resource abundance is highest in the centre of the environment, at $x=0$. In particular, $R_0(x)$ is given by the following bell-shaped curve (for $-\Omega \leq x \leq \Omega$):

$$R_0(x) = R_0(0) \exp\left(\frac{-x^2}{\beta_R}\right).$$ (1)

Thus, $R_0(0)$ gives the value of R_0 at $x=0$, and β_R determines the rate at which the input of resources falls off as we move away from $x=0$. We assume $R_0(x)=0$ for $x < -\Omega$ and for $x > \Omega$.

During each generation, both the sexual and the asexual species simultaneously undergo a birth stage, followed by the death of the adults. After that, some of the juveniles die due to viability selection. Next, the remaining juveniles migrate to new locations within the habitat. Finally, the remaining juveniles mature into adults and produce the next generation.

During the birth phase, each adult at a given location x consumes resources at a rate $\gamma R(x)$, where $\gamma > 0$. Thus, the rate at which individuals capture resources is assumed to be lower when resources are scarce than when they are abundant. Let $N_S(x)$ and $N_A(x)$ represent the density at location x of adults of the sexual and asexual species, respectively. Under the foregoing assumptions, we can show that the amount of resources consumed by an individual declines as $R_0(x)$ decreases, and as the local density of sexual and asexual individuals increases (the exact relations are given in the Appendix).

The expected number of offspring produced by an adult member of either species is assumed to be proportional to the amount of resources that the adult consumes (for sexuals, this number does not include offspring produced by male effort - i.e., via donation of sperm or pollen). For convenience, and without loss of generality, say that resources are scaled so that a member of the sexual species produces, on average, one juvenile for every unit of resources that she consumes, while the average number of offspring produced by members of the asexual species is assumed to be Θ for every unit of resources consumed (where $\Theta > 0$). Thus, if $\Theta < 1$, then asexuals are less efficient than sexuals, while the reverse holds if $\Theta > 1$.

We consider a single phenotypic character, z (where $-\infty < z < \infty$). We assume that the value of z for any individual is determined entirely by the individual's genotype. We ignore stochastic environmentally induced variation among individuals living in the same location because, under typical assumptions, such variance simply diminishes the strength of selection (Turelli 1984). We study a range of strengths of selection (see below) and thus, incorporation of environmental variance would not qualitatively alter the results.

Say that a substantial number of loci act together in an additive manner to determine the value of z. To simulate this situation in an approximate (but manageable) way, we use the "infinitesimal model (Bulmer 1980) "and assume that, for any given member of the sexual species, the value of z is equal to $\dfrac{z_M + z_F}{2} + \Psi$, where z_M and z_F are the values of z for the individual's mother and father. The value of Ψ represents variation introduced as a result of recombination and segregation. We assume that Ψ is distributed according to a "discretised normal distribution", with mean zero and approximate standard deviation denoted by σ (see the Appendix for details). The value of z for an asexual individual is given by z_M, where z_M is the value of z for the individual's one and only parent. Thus, asexuals are genetically identical to their parents.

We assume that adult members of the sexual species choose their mates at random from among the adults in their immediate vicinity (i.e, from among adults who share the same value of x). Each adult chooses a new mate for each offspring she produces.

Note that we can expect a highly-selfed organism to be almost entirely homozygous, and thus there will be very little genetic variance among the offspring of any given parent. This implies that our model of asexuality can be used to represent a highly-selfed population, and thus the results of the model apply to selfers, as well as to asexuals.

After birth, viability selection occurs. At any given location x, the probability that an individual with phenotype z will survive viability selection is denoted by $W(x,z)$, and it is given by the following "nor-optimal" function:

$$W(x,z) \;=\; \exp\left(\frac{-(z-x)^2}{\beta_V} \right) \qquad (2)$$

where $\beta_V > 0$. Thus, the optimum phenotype at any given location x is given by $z = x$. The rate at which viability falls off with the distance of a phenotype from the local optimum (i.e., the strength of selection) is determined by the magnitude of β_V. Equation 2 holds for $-\Omega \le z \le \Omega$. We assume that, for values of z outside of this range, $W(x,z)=0$. This last assumption was made for technical reasons, and we showed that it is unlikely to have any qualitative effect on the results by analysis of special trials in which Ω was set to exceptionally large values.

After viability selection, juveniles are assumed to migrate (these "juveniles" may be seeds, in the case of a plant). Each juvenile is equally likely to move east or west. Regardless of which direction they move, the distance that they move is based on an exponential distribution. Further details of the migration scheme are given in the Appendix.

3. Results

To investigate the model, we ran 1,000 computer simulation trials. Parameter values were chosen at random, as were the initial frequencies of the various phenotypes at each location in space (the exact procedures are described in the Appendix). Both the sexual species and the asexual species were present in the environment at the start of each trial. Each trial was run until the average density of both species was changing by less than 10^{-6} % over the course of 100 generations. (The average density of the sexual species is given by $\sum_{x=-\Omega}^{\Omega} P_S(x)N_S(x)$, where $P_S(x)$ represents the proportion of sexual adults at location x. The value of $P_S(x)$ is given by $N_S(x)/\sum_{i=-\Omega}^{\Omega} N_S(i)$. Similarly, $\sum_{x=-\Omega}^{\Omega} P_A(x)N_S(x)$ gives the average density of asexuals, where $P_A(x)$ represents the proportion of asexuals at location x.)

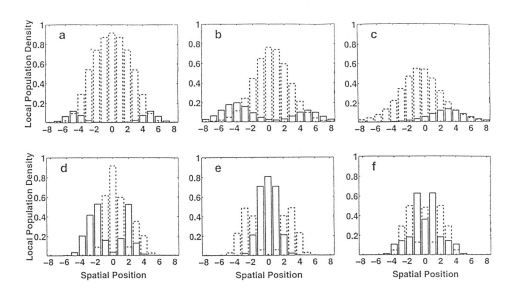

FIGURE. Equilibrium density distributions from a variety of simulation runs. The dotted lines show the density of the sexual species (the $N_S(x)$ values) and the solids lines are for the asexuals (the $N_A(x)$ values). In 77.2% of the polymorphic trials the results were similar to those shown in panels a and b, in that distribution for the sexuals had one central peak, while the distribution for the asexuals had two peaks, one on either side of the peak in the sexual species' distribution. Panels a-d show the outcome of four trials that were collected during the main set of 1000 simulation runs. The runs shown in panels e and f used the same parameter values as the run shown in panel d, but the initial density distributions were different (as explained in the text).

In 1.4% of trials both species went extinct at the end of the trial, or were at a very low (and declining) density. This tended to occur when γ was very small, so that resource consumption was extremely inefficient. In 41.6% of the trials, only the sexuals established a stable density distribution, while asexuals went extinct, or were on their way to extinction (an average density less than 0.1% of that for the sexuals at the end of the trial, and falling). This happened only when $\Theta < 1$, so that the sexuals had an intrinsic fertility advantage. In 25.4% of the trials the asexuals took over (i.e., only the asexuals established a stable density distribution, while sexuals went extinct or were on their way to extinction). Excepting the cases where both species went extinct, take-over by the asexuals happened whenever $\Theta > 1$, and in some cases when $\Theta < 1$. In all cases where only one species had a substantial density at the end of the trial, this remaining species established a symmetric (or nearly symmetric) bell-shaped density distribution.

In 31.6% of the trials, both species had a substantial frequency at equilibrium (i.e., at the end of the trial). We call these the *polymorphic trials*. In all of the polymorphic trials, the sexuals had an intrinsic fertility advantage ($\Theta < 1$). In every polymorphic trial, the equilibrium distributions showed that members of the asexual species were, on average, further from the resource-rich centre of the environment than were members of the sexual species. That is to say, in the polymorphic trials, at equilibrium, we had $\displaystyle\sum_{x=-\Omega}^{\Omega} P_A(x) \mid x \mid > \sum_{x=-\Omega}^{\Omega} P_S(x) \mid x \mid$. Thus, the only pattern seen in these trials was asex outside.

In 77.2% of the polymorphic trials, the equilibrium density distribution for the sexual species had a single peak, and the equilibrium distribution for the asexuals had two peaks, one on either side of the peak for the sexuals (see Figures 1a and 1b). In 22.5% of the polymorphic trials, the equilibrium density distributions for both the sexual species and the asexual species had a single peak (see Figure 1c). In these cases, the density distributions for both species tended not to be symmetric about the centre of the environment. In one trial (0.3% of all the polymorphic trials) the density distributions for both species had two peaks (as shown in Figure 1d).

We used variance in spatial position to measure species range. Thus, for example, the range of the sexual species is given by,

$$\sum_{x=-\Omega}^{\Omega} P_S(x)(x - \bar{x}_S)^2 \text{ ,where } \bar{x}_S = \sum_{x=-\Omega}^{\Omega} P_S(x)x \text{ gives the mean position of sexuals.}$$

We found that, in 82.3% of the polymorphic trials, the sexuals had a smaller range at equilibrium than did the asexuals. Furthermore, in all of the polymorphic trials where the asexual equilibrium density distribution had two peaks, the range for the asexuals was wider than the range for the sexuals. However, at equilibrium, the asexuals had a wider range in only 21.1% of the cases where the equilibrium density distributions of both species had only a single peak.

One interesting aspect of the model is that the equilibrium reached is highly dependent on the initial conditions. We found that a given set of parameter values could sometimes produce a variety of different outcomes, depending on the initial spatial distributions of the various phenotypes within the sexual and asexual species. To demonstrate this property of the model, we used a trial using the parameter values that were randomly selected during the series of 1000 trials, and that led to the equilibrium distributions shown in Figure 1d. We re-ran this trial, but instead of picking the initial density distributions at random, we selected distributions that gave asexuals a large initial advantage near the centre of the environment, and that gave the sexual species a large initial disadvantage in the same region. The result, at equilibrium, is shown in Figure 1e. The figure shows that, in this case, the asexuals are, on average, closer to the resource-rich centre of the environment than are the sexuals. Other manipulations of the initial density distributions allowed for a variety of different outcomes. For example, Figure 1f shows a case where the equilibrium distribution of the sexual species has three peaks. This trial used the same parameter values that were used to produce Figures 1d and 1e.

4. Discussion

The results suggest that asex outside is a pattern that tends to arise naturally whenever sexual and asexual species coexist. There also seems to be a natural tendency for asexual species to have wider geographic ranges than their outcrossing competitors. Furthermore, the results suggest that both species can be maintained in a stable polymorphism only if there is a fertility disadvantage associated with asexuality ($\Theta < 1$). Such a disadvantage might arise from one of the well-known mechanisms that have been proposed for the evolution of sex (e.g., Muller's Ratchet (Charlesworth *et al.* 1993; Muller 1964; Pamilo *et al.* 1987; Peck *et al.* 1997) or parasite evasion (Hamilton *et al.* 1990)). Finally, the results suggest that the history of the settlement of a habitat can have a strong effect on the spatial distribution of species within that habitat. This "history effect" can be sufficiently strong to overpower the tendency to achieve asex outside. Because of the homozygosity induced by selfing, these same generalisations apply to selfing taxa that are in competition with related oubreeders.

Why does the model produce the results that it does? To answer this question, it is useful to consider a sexual population that has inhabited the environment described above for a large number of generations, in the absence of any competitor species. The sexual population will achieve a symmetric density distribution, with its highest density at $x=0$ (where resource-input is high) and lowest density towards the ends of the habitat. Because the density is relatively high near the centre of the habitat, this part of the habitat receives a relatively small number of migrants, when compared to the non-migrant population. However, a very different situation prevails at the edges of the habitat. In comparison to locations close to the centre, a large proportion of individuals near the edges are migrants. Most of these have moved from more central regions. This is not because individuals near the centre are more likely to migrate, but simply because there are so many of them. As a result of the relatively large proportion of migrants present at the edges of the habitat, the local populations in these areas tend to suffer from the presence of locally unfit genotypes (i.e., from "migration load"). That is, many of the potential mates on the edges are better adapted to the environment near the centre of the habitat than they are to the habitat in which they find themselves. This is because they are migrants, or are recently descended from a migrant.

Because of the high level of migration load on the edges of the habitat, it may be possible for an asexual species to out-compete the sexual species in these marginal regions. However, if the asexual species has an intrinsic fertility disadvantage, then it is unlikely to spread to the centre of the habitat, where the sexual species does not suffer greatly from migration load. Note that "migration load" is a well-known phenomenon in natural populations (Antonovics 1968; Dias 1996) and it has recently been implicated in the establishment of the limits of species ranges (Kirkpatrick and Barton 1997).

We used our computer model to carry out some additional investigations. We found that, if we replace the asexual species with a second sexual species (for which Θ indicates its relative advantage or disadvantage, as before) then coexistence is possible, but if $\Theta < 1$, then the second species tends to accumulate on the edges of the habitat, whereas, if $\Theta > 1$, the second species tends to accumulate in the centre, and the first species accumulates on the edges. Thus, whether the second species is sexual, asexual, or highly selfed, if it is intrinsically weaker than the first species, then it will tend to accumulate in the resource-poor margins of the habitat. This may be an important ecological principle.

Our additional investigations also showed that the mixing of lineages inherent in outbreeding is important for the coexistence of species. In particular, we found that, when we competed two asexual species against each other (arbitrarily numbered as species 1 and species 2) species 1 tended to displace species 2 throughout the habitat when $\Theta < 1$, and species 2 tended to displace species 1 when $\Theta > 1$ (where $\Theta > 1$ means that species 2 has an intrinsic advantage, and vice versa when $\Theta < 1$). This occurred even if we modified the model by allowing variation in genotypic values among siblings, which makes the model more similar to the model of sexual reproduction via outbreeding.

In addition to occurring in resource-poor habitats, asexual species also tend to occur further from the equator than their sexual competitors, and at higher altitudes (Bell 1982; Bierzychudek 1985; Glesener and Tilman 1978; Hughes 1989; Vandel 1928). Highly selfed plants also tend to occur far from the equator (Lloyd 1980). We suggest that these patterns may be explained by a process similar to the one described here. Typically, high latitudes and high altitudes are associated with short growing seasons. This can leave outbred sexual populations in these areas extraordinarily susceptible to immigration from more equatorial areas (or from lower altitudes) where growing seasons are longer (Lindgren et al. 1995). Asexuality and selfing can provide defences from the resulting migration load by eliminating the possibility of combining locally adapted genotypes with maladapted migrant genotypes. These ideas are explored elsewhere (Peck et al. 1998).

The processes we have described are not the only possible explanations for geographic patterns found in asexual and in self-fertilising taxa (Jarne and Charlesworth 1993; Lynch 1984; Parker 1998). For example, geographic parthenogenesis may occur because, in marginal habitats, finding a mate is costly (Bierzychudek 1985; Lynch 1984) . A similar hypothesis can be used to explain the occurrence of self-fertile taxa in marginal habitats (Jarne and Charlesworth 1993). However, this "reproductive assurance" hypothesis seems unlikely in many animals (Bell 1982), and it may not account for the occurrence of asexuality and selfing at high latitudes, where populations are often dense (Peck et al. 1998). It may be the case that reproductive assurance is important in some cases, while the avoidance of migration load (or some other process) is important in others. Our hope is that the results presented here will assist in the design of the empirical studies that will, eventually, explain the geographic distributions of asexuality and self-fertilisation.

5. References

Antonovics, J., 1968 Evolution in closely adjacent plant populations. V. Evolution of of self-fertilty. *Heredity* **23**: 219-238.

Barrett, S. C. H., 1988 The evolution, maintenance, and loss of self-incompatibility systems. In *Plant Reproductive Ecology*, edited by J. L. Doust and L. L. Doust. Oxford University Press.

Bell, G., 1982 *The Masterpiece of Nature*. University of California Press, San Francisco.

Bierzychudek, P.,1985 Patterns in plant parthenogenesis. *Experientia* **41**: 1255-1263.

Bulmer, M., 1980 *The Mathematical Theory of Quantitative Genetics*. Clarendon Press, Oxford.

Charlesworth, D., M. T. Morgan and B. Charlesworth, 1993 Mutation accumulation in finite outbreeding and inbreeding populations. *Genetical Research* **61**: 39-56.

Dias, P. C., 1996 Sources and sinks in population biology. *Trends in Ecology and Evolution* **11**: 326-330.

Glesener, R. R. and D. Tilman, 1978 Sexuality and the components of environmental uncertainty: Clues from geographic parthenogenesis in terrestrial animals. *The American Naturalist* **112**: 659-673.

Hamilton, W. D., R. Axelrod and R. Tanese, 1990 Sexual reproduction as an adaptation to resist parasites (a review). *Proceedings of the National Academy of Sciences (USA)* **87**: 3566-3573.

Hughes, R. N., 1989 *A Functional Biology of Clonal Animals*. Chapman and Hall, London.

Jarne, P. and D. Charlesworth, 1993 The evolution of the selfing rate in functionally hermaphrodite plants and animals. *Annual Review of Ecology and Systematics* **24**: 441-466.

Kirkpatrick, M. and N. H. Barton, 1997 Evolution of a species range. *The American Naturalist* In press.

Lindgren, D., L. Paule, S. Xihuan, R. Yazdani, U. Segerström, W. Jan-Erik and L. Dejdebro, 1995 Can viable pollen carry Scots pine genes over long distances? *Grana* **34**: 64-69.

Lloyd, D. G., 1980 Ecological and geographical distribution of self- and cross-fertilization. *In Demography and Evolution in Plant Populations*, edited by O. T. Solbrig. Blackwell, Oxford.

Lynch, M., 1984 Destabilizing hybridization, general-purpose genotypes and geographic parthenogenesis. *The Quarterly Review of Biology* **59**: 257-290.

Maynard Smith, J., 1978   *The Evolution of Sex.*   Cambridge University Press, Cambridge.

Muller, H. J., 1964 The relation of recombination to mutational advance. *Mutation Research* **1**: 2-9.

Pamilo, P., M. Nei and W. H. Li, 1987 Accumulation of mutations in sexual and asexual populations. *Genetical Research* **49**: 135-146.

Parker, E. D., 1998 Genetic structure and evolution in parthenogenetic animals. In *Evolutionary Genetics from Molecules to Morphology*, edited by R. Singh and C. Krimbas. Cambridge University Press, Cambridge.

Peck, J. R., G. Barreau and S. C. Heath, 1997 Imperfect genes, Fisherian mutation and the evolution of sex. *Genetics* **145**: 1171-1199.

Peck, J. R., J. M. Yearsley and D. Waxman, 1998   Explaining the geographic distributions of sexual and asexual population. *Nature* **391**: 889-892.

Suomalainen, E., 1950 Parthenogenesis in animals. *Advances in Genetics* **3**: 193-253.

Turelli, M., 1984 Heritable genetic variation via mutation-selection balance: Lerch's zeta meets the abdominal bristle. *Theoretical Population Biology* **25**: 138-193.

Vandel, A., 1928 La parthéngénèse géographique: contribution l'étude biologique et cytologique de la parthénogénèse naturelle. *Bulletin Biologique de la France et de la Belgique* **62**: 164-281.

6. Appendix

In the main simulation trials we set $\Omega=8$, so that there were 17 locations considered in the habitat. We also allowed for 17 possible values of z, one optimal for each location (i.e., z took on values -8, -7, -6, ... 8). To initialise each trial, we chose 17 random numbers using a uniform distribution on the interval [0, 1/17]. These were assigned as the initial densities of sexuals at location $x=0$ for z values -8, -7, -6, ... 8. Exactly the same procedure was used to get the initial densities for the asexual species for $x=0$. These procedures were then repeated to get initial densities for the two species for all non-zero values of x that satisfied $-8 \leq x \leq 8$.

If the average density for either species fell below 10^{-3} at any point during a trial, that species was regarded as extinct, and its local density was set to zero for all allowable values of x.

The value of Ψ is distributed according to a "discretised normal distribution."

If the mean value of a pair of parents is given by $\bar{z}$, then the probability that the phenotype of a particular offspring takes on an integer value i ($V(i\mid\bar{z})$) is given by:

$$V(i\mid\bar{z}) \;=\; Q\exp\!\left(\frac{-(i-\bar{z})^2}{2\sigma^2}\right), \qquad (A1)$$

where Q is a normalization factor chosen to ensure that $\sum_{i=-\infty}^{\infty} V(i \mid \bar{z}) = 1$.

We set $R_0(0)=1$ for all simulation studies. Other parameters were chosen at random for each trial from rectangular distributions. The intervals for these rectangular distributions were selected so as to ensure a substantial number of trials led to the coexistence of both species. The intervals selected were as follows: For γ, [2, 20]. For Θ, [0.7, 1.1]. For β_R, [6, 36]. For β_m, [0.1, 3]. For β_V [1.0, 6.0]. And for σ, $[7 \times 10^{-11}, 0.5]$.

From the assumptions about resource consumption, we have:

$$\frac{dR(x)}{dt} = -\gamma R(x) \left[N_S(x) + N_A(x) \right]. \tag{A2}$$

Without loss of generality, we assume that the birth phase is a single time unit in duration. With this assumption, the amount of resources extracted from the environment by each adult during the birth phase at location x is given by:

$$\frac{R_0(x)\left(1 - \exp\left(-\gamma \left[N_S(x) + N_A(x)\right]\right)\right)}{N_S(x) + N_A(x)} \tag{A3}$$

During the migration phase, the probability that an individual at location i migrates to location j is given by:

$$\tanh\left(\frac{1}{2\beta_m}\right)\exp\left(\frac{-|i - j|}{\beta_m}\right). \tag{A4}$$

Thus, the probability that an individual remains at the location where it was born is given by $\tanh\left(1/2\beta_m\right)$. The average absolute distance that an individual moves is $1/\sinh(\beta_m)$. The probability that an individual at $x=i$ will migrate beyond $x=-\Omega$ or $x=\Omega$ (i.e., out of the habitat) is equal to $\exp\left(-[2\Omega+1]/(2\beta_m)\right)\cosh\left(i/\beta_m\right)/\cosh\left(1/(2\beta_m)\right)$. As $R_0(x)=0$ outside the habitat, these migrants do not produce offspring.

Centre for the Study of Evolution
The University of Sussex
Brighton BN1 9QG
United Kingdom

E-mail: j.r.peck@sussex.ac.uk, nhi785@abdn.ac.uk, dwaxman@sussex.ac.uk

Received: 30 March 1998

Lectures on Mathematics in the Life Sciences
Volume **26**, 1999

The Social Life of Automata

Karl Sigmund

ABSTRACT. This is a survey of recent results on the evolutionary dynamics of cooperation. Based on the Prisoner's Dilemma Game, evolutionary chronicles show the emergence of cooperative, error-proof strategies like Pavlov or Contrite Tit For Tat. Such strategies describe social norms. Another class of models deals with indirect reciprocity: here, the return of an altruistic action can be due to a third party. Again, strategies based on discrimination can evolve and lead to cooperation.

1. Introduction

Some of the major transitions in evolution occurred only once, and others several times (*cf.* Maynard Smith and Szathmáry, 1995). Eusocial colonies, for instance, emerged repeatedly among bees, ants, termites, and aphids. This allows one to compare the importance of different factors, for instance ecological opportunities, life history traits, genetical structures etc. and even to predict where similar societies are likely to be found (as Richard Alexander did in anticipating the sterile worker caste of the naked mole-rat, see Sigmund, 1995). In contrast, no parallel to human societies is known in the history of evolution; we seem to be unique in having achieved a social structure distinguished (i) by the levelling of reproductive opportunities, (ii) by the prevalence of division of labour, mutual help and economic exchange between non-related individuals, (iii) by information transfer based on language and (iv) by moral obligations both externally enforced through group sanctions and internalized through powerful emotions.

We can, of course, learn much about the cultural determinants of human societies by comparing tribes, clans, states and gangs; but all these are manifestations of a universal 'human nature' caused by a major biological transition which, apparently, has occurred once only. In order to analyze the mechanisms responsible for it, we have to use thought experiments.

Game theory was devised explicitly as a tool for the social sciences. It was meant to model the independent decision-making process of interacting individuals, each bent upon the 'selfish' goal of maximizing his or her own payoff. Interpreting payoff as reproductive fitness provides a good tool for studying individual selection based on Darwinian competition. But for decades, game theory was handicapped

1991 *Mathematics Subject Classification.* 92H10 and 90D80.

by the fiction of the 'rational player', despite the fact that many economists, and every psychologist, knew better. Furthermore, it was only when biologists started to use game theory that populations of individuals were considered (see Maynard Smith, 1982, Binmore, 1992, Weibull, 1995, Hofbauer and Sigmund, 1998).

The advent of evolutionary game theory has changed all this. Individual players were no longer assumed to be rational, but to follow simple, knee-jerk rules. In the spirit of Richard Dawkins (1976), who claims that we are mere robots, players were therefore modelled by simple automata. Populations of such interacting automata, engaged in the massively parallel kind of problem-solving characteristic of Darwinian evolution, were studied by means of computer simulations or mathematical analysis following their evolution for many generations. New strategic variants (or programs) were introduced either by random processes or by hand, and tested against the current composition of the population. Nonlinear dynamics describing the resulting adaptation or selection processes, were used to analyze the chronicles of these artificial societies (see, *e.g.*, Axelrod, 1997).

This program of evolutionary game theory has been applied to a wide variety of biological and (more recently) economic topics. In this paper, we sketch some recent developments in one particularly active field: the evolution of cooperation. More precisely, we deal with one of the three factors currently recognized as essential, namely with reciprocity. This is not meant to downplay the importance of the other two factors, *viz.* relatedness (Hamilton, 1963) and group selection (Wilson and Sober, 1994). Doubtlessly close kinship ties, the major basis of cooperation in clones and bee colonies, did prevail in early hominid groups. Furthermore, group selection – more precisely, the individual selective advantage due to belonging to a successful group – was essential, because the major threats to survival were most likely coming from rival groups. This being said, let us turn to the subject of reciprocal altruism, originally introduced in a landmark paper by Trivers (1971) which, to this day, serves as an inspiration to the field.

2. Reciprocal Altruism and the Prisoner's Dilemma

Assume that in an encounter between two players, one is a potential donor and the other a recipient. The donor can give help that the recipient needs. Giving help costs $-c$ to the donor and yields payoff b to the recipient (with the payoff interpreted as Darwinian fitness, *i.e.* reproductive success, and assuming $0 < c < b$). According to Hamilton's rule, it will pay to help if the degree of relatedness r between donor and recipient (*i.e.* the probability that a randomly chosen gene from the donor also belongs to the recipient's genome) is larger than the cost-to-benefit ration, *i.e.*

$$r > c/b.$$

(Since $r \leq 1/2$ under normal circumstances – barring identical twins or high inbreeding – this condition requires $c < b/2$). Does this imply that one should never help an unrelated individual? Not so, according to Trivers, if there is a reasonable chance that the recipient is able to return the help. This is the principle of reciprocal altruism – in Trivers' definition, 'the trading of altruistic acts in which benefit is larger than cost, so that over a period of time both enjoy a net gain' (Trivers, 1985, p. 361). Accordingly, 'reciprocal altruism is expected to evolve when two individuals associate long enough *to exchange roles frequently* as potential altruist and recipient'.

This, however, opens the door to unilateral defection. Suppose that in two consecutive rounds, the players exchange the roles of donor and recipient. If both help each other, both obtain $b - c$. This is higher than the payoff 0 obtained if both refrain from helping. But is one helps and the other does not, then the helper is left with the costs of his act, $-c$, and the recipient gets away with b points. This is just the rank ordering of the payoff values for the Prisoner's Dilemma, which had been studied for many years by game theorists and experimental psychologists. The difference is merely that both players have to decide simultaneously whether to cooperate (play **C**) or to defect (play **D**). If both play **C**, both receive the reward R for mutual cooperation; if both play **D**, both receive the punishment P; and if one defects unilaterally, he receives the temptation T whereas the other player is left with the sucker's payoff S. For the Prisoner's Dilemma game, it is assumed that

$$T > R > P > S \quad \text{and} \quad 2R > T + S.$$

The first condition implies that the dominant option is to play **D**: it yields a higher payoff, no matter what the other player is choosing. In the donor-recipient game (with two rounds in alternating roles), one has $T = b, R = b - c, P = 0$ and $S = -c$, so that these inequalities are trivially satisfied.

The conclusion seems inescapable that if the interaction is not repeated, co-operation cannot emerge. It turns out, rather surprisingly, that this conclusion is premature. But we will return to this point only in the concluding discussion – our aim here is rather to follow Trivers in assuming that individuals experience, on average, several interactions.

Let us suppose, then, that the game is repeated with a constant probability w. The number of rounds is a random variable with expected value $(1 - w)^{-1}$. The total payoff is given by $\sum A_n w^n$, with A_n as payoff in the n-th round and w^n the probability for the n-th round to occur. In the limiting case $w = 1$ (the infinitely iterated game) one uses as payoff the limit in the mean, i.e. $(A_1 + ... + A_n)/n$ (provided it exists). If w is sufficiently large, there exists (in contrast to the one-shot game) no strategy which is best against all comers (see Axelrod, 1984). For $w > (T - R)/(T - P)$, for instance, the best reply against *Always*C is to always defect, whereas against *Grim* (the strategy that cooperates up to the first time that it is been exploited, and from then onwards always defects) it is best to always cooperate.

In a series of round robin tournaments, Axelrod found that the simplest strategy submitted, namely TFT, finished first. Furthermore, Axelrod and Hamilton (1981) explored the emergence of cooperation in evolving populations of players. In particular, they showed that the two strategies *Always*D and TFT (the Tit For Tat strategy that cooperates in the first round and from then on always repeats the previous move of the co-player) are in bistable equilibrium: none of them can invade the other. But as soon as the frequency of TFT players exceeds a certain threshold (given by $c(1 - w)/w(b - c)$ in the donors-recipient game), keeps growing. For large values of w, this threshold is very small. This means that a small cluster of TFT players can invade a population of defectors: the few interactions with their like more than compensate their loss against the resident majority of defectors.

One may not conclude, however, that a population dominated by TFT can resist invasion by all comers. This becomes particularly obvious if we take into account the possibility for errors, which must always be present in realistic situations. In fact, the interaction between two TFT players is particularly sensitive to noise. One wrong move causes a whole chain of alternating defections. One further mistake can lead back to mutual cooperation, but just as well to mutual defection. The average payoff decreases drastically. Obviously, the two players should be able to forgive occasionally – not according to a regular pattern, for this could be exploited, but rather on a random basis.

This leads to stochastic strategies. We often do not use hard and fast rules in our everyday interactions, but are guided by factors which are difficult to predict, and which result in a stronger or weaker propensity to opt for this move or that (May, 1987). With TFT, this propensity is 100 percent or 0 percent, depending on whether the co-player cooperated in the previous round or not. With an error rate of 1 percent, TFT cooperates with 99 or with 1 percent probability.

3. Evolutionary Chronicles

Let us consider this in a more general setting. In each round, there are 4 possible outcomes, leading to 4 different payoff values. If we assume that each outcome determines the next move of the player, this yields 16 different strategies (32 if we include the first move). If we allow in addition stochastic strategies with a larger or smaller propensity to cooperate, we obtain a 4-dimensional space of strategies given by quadruples (p_R, p_S, p_T, p_P) where p_i is the probability to play **C** after outcome i. We can use a computer to find the most successful strategy, by introducing occasionally a small minority of a new, randomly chosen strategy into the population and watching how its frequency develops under the influence of selection. If we run this for a sufficiently long time, we can test a large sample of strategies.

Such mutation-selection chronicles depend on contingencies, and can take very different paths, but they frequently lead to a population dominated by the so-called Pavlov strategy $(1, 0, 0, 1)$ – the strategy that cooperates if and only if, in the previous round, the co-player used the same move as oneself (Nowak and Sigmund, 1993). This strategy embodies a simple win-stay, lose-shift rule: it repeats the former move if the payoff was high (T or R) and switches to the alternative move if the payoff was low (*i.e.* P or S). Arguably, this is the simplest learning rule. Pavlov seems a hopeless strategy for invading an *Always*D population, since it gets suckered every second round. In fact, it needs a retaliatory strategy like TFT or *Grim* (the strategy that cooperates until suckered, and from then on never cooperates). Once such 'nice' strategies (strategies that are not the first to defect) have taken over, Pavlov can invade, because it is tolerant to errors. If two Pavlov players are engaged in a repeated PD game, and one of them commits a mistake and defects, then both players will defect in the next round – the 'sinner' because he is happy with his T and repeats the former move, the 'sucker' because he shifts to the other option. As a result, both players obtain the low payoff P, switch again and thereby resume mutual cooperation. In addition to being tolerant to errors, Pavlov has also the advantage of being intolerant to *Always*C players. After a mistaken defection against such a player, Pavlov keeps defecting. Therefore, indiscriminate altruists cannot spread in a Pavlov population (whereas they could spread by neutral drift in an TFT population). As a result, defectors find no easy victims. They only find Pavlov-players, whom they can exploit in every second round (obtaining $(T + P)/2$ in the mean), whereas Pavlov players obtain R against each other. As long as $2R > T + P$ – or, in the case of the donor-recipient game, as long as $c < b/2$ – a Pavlov population is stable against invasion.

4. Social Norms

How can one formulate this kind of stability? The usual approach would be to test for evolutionarily stability (*cf.* Maynard Smith, 1982). It can easily be shown, however, that in the context of repeated games, such strategies do not exist. In particular, neither TFT nor *Always*D are evolutionarily stable, although this has occasionally been claimed. Possibly the most appropriate notion in this context is that of a limit evolutionarily stable strategy. This was originally formulated in terms of extensive games (Selten, 1975), but for repeated games, it is more appropriate to formulate it in terms of strategies implemented by finite automata (Leimar, 1997). In each round, such an automaton can be in one of m internal states. Depending on the state, it plays **C** or **D**; and depending on the outcome of this round, it switches to the next state. In order to test whether such a strategy is a limit ESS, one assumes that it plays against a copy of itself, and looks at what happens at every outcome. If the sequence of moves prescribed by the strategy is better than any alternative, then the strategy is a limit ESS: it is always disadvantageous to deviate from it. In fact, a limit ESS is a social norm: if everybody adheres to it, it wouldn't do not to do.

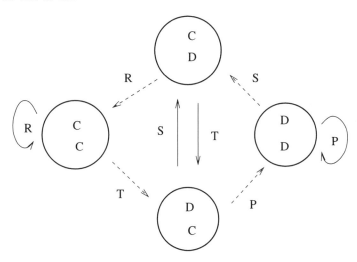

FIGURE 1. TFT is not a norm.

Let us consider this for TFT. We may view the internal state as defined by the outcome R, S, T or P of the previous round. The action rule is to play **C** after R and T. Fig. 1 shows a graph whose vertices are the four states (the move on top is that by the player, the move below that by the co-player). The full arrow shows the transition to the next state, if the player uses the move prescribed by TFT; the broken arrow shows the transition if the player uses the alternative move. This holds under the assumption that the other player sticks to TFT. We see that in the state S, it would be better to follow the broken arrow: the payoff for the next two moves is $2R$, which is larger than $T + S$. We should note that in principle, the state S should never be reached in a game between two TFT players. But according to the *trembling hand* doctrine of Selten (1975), it can be reached if one player misimplements his move, a mistake which may happen with a small, but positive probability.

If we study the same situation for Pavlov, we obtain the graph of Fig. 2 (again assuming the co-player to use Pavlov). If $2R > T + P$ it is best to follow the full arrows, *i.e.* Pavlov is a limit ESS. If $2R < T + P$, it is better, when in state P, to deviate from the Pavlov rule and play **D**. In this case, Pavlov is not a limit ESS.

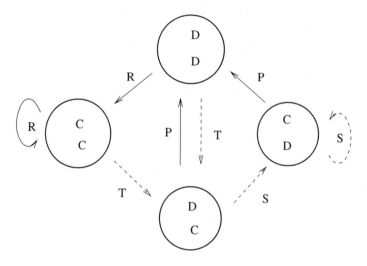

FIGURE 2. Pavlov is a norm if $2R > T + P$.

It is easy to find automata that are always norms and lead to cooperation. Let us consider the following example, which has three states and starts in state 1. In state 1 or 2, it plays **C**, and in state 3 it plays **D**. The transition table is given by Table 1. Again, it is easy to check that it is best to always follow the full arrow.

	R	S	T	P
1	1	2	3	1
2	1	2	1	2
3	1	1	3	3

Table 1

This strategy becomes very transparent if one interprets it as Contrite Tit For Tat (cTFT, originally introduced by Sugden, 1986). It is based on the notion of a *standing* associated to each player, which can be g (good) or b (bad). In each round, the player acts (*i.e.* opts for **C** or **D**) and obtains a new standing which depends on that action and on the previous standing of both players. The rules for updating the standing are the following: if the co-player has been in good standing, or if both have been in bad standing, one receives a good standing if one cooperates, and a bad standing otherwise. If one has been in good standing and the co-player in bad standing, one receives a good standing no matter what one does.

Thus if one cooperates in a given round, one will always obtain a good standing: but if one defects, one will be in good standing only if the defection has been 'provoked' – *i.e.* if one has been in good standing and the opponent in bad standing.

cTFT is the strategy which cooperates except if the player is in good standing and the co-player is not. This means that the player defects when provoked, but not

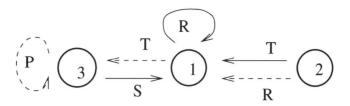

FIGURE 3. cTFT is a norm.

otherwise. A player who defects by mistake knows that he lost his good standing, and meekly accepts punishment, *i.e.* keeps cooperating even if the other player uses **D** on him.

In other words, cTFT begins with a cooperative move, and cooperates except if provoked (or by mistake). If two players using this strategy engage in a repeated Prisoner's Dilemma, and one player defects by mistake, then he loses his good standing. In the next round, he will cooperate, whereas the other player will defect without losing his good standing. From then on both players will be in good standing again and resume their mutual cooperation in the following round.

This is exactly the strategy implemented by the previously described automaton: the state 1 corresponds to both player being in the same standing, the state 2 occurs when the player is in good standing and the co-player not, and state 3 is just the mirror image of 2. Hence cTFT is a norm, and therefore uninvadable. Moreover, it is itself as adept at invading a population of defectors as TFT: it can only be suckered in the first round, and retaliates from then on. Moreover, it is immune against mistakes of implementation.

However, in contrast to Pavlov, cTFT is vulnerable to errors in perception (*cf.* Boerlijst *et al.*, 1997). A player erroneously believing to have been suckered will play **D**. From then on, both cTFT players will remain in state 2 and keep punishing their co-player in good faith. Another weak point of cTFT is that it does not exploit *Always***C**-players, who therefore can spread by neutral drift and thereby open the door to defectors. In Sigmund *et al.* (1998) it is argued that the investigation of strategies implemented by automata should take account of mistakes in perception just as well as of mistakes in implementation. Given that the number of norms is extremely large (see Lindgren, 1991, and Leimar, 1997), it is difficult to predict which strategies will eventually be selected.

5. Indirect Reciprocity

So far, we have described direct reciprocation. Alexander has suggested that another, indirect reciprocation is also operating in human societies, and that it is the basis of all moral systems. In fact, indirect reciprocity was mentioned by Trivers (1971) as 'generalised altruism': the return of an altruistic act is directed towards a third party. 'Individuals ... may respond to an altruistic act that benefits themselves by acting altruistically toward a third individual uninvolved in the initial interaction... In a system of strong multiparty interactions, it is possible that in some situations individuals are selected to demonstrate generalised altruistic tendencies.' This possibility is further stressed in Trivers (1985), who explores the possibility that a sense of fairness may evolve 'in species such as ours in which a system of multi-party altruism may operate such that an individual does not necessarily receive reciprocal benefit from the individual aided but may receive the return from third parties.'

Richard Alexander extended this idea under the heading of 'indirect reciprocity' (see Alexander, 1979 and 1986). With indirect reciprocity, one does not expect a return from the recipient (as with direct reciprocity), but from someone else. Cooperation is thereby channeled towards the cooperative members of the community. A donor provides help if the recipient is likely to help others, or at least if he has not been observed withholding help. According to Richard Alexander (1986), indirect reciprocity, which 'involves reputation and status, and results in everyone in the group continually being assessed and reassessed', plays an essential role in human societies. Alexander argues that systems of indirect reciprocity are the basis of moral systems. (For a dissenting opinion, see the message from the Pope, 1997).

Such scenarios have been considered by game theorists. In Boyd and Richerson (1989) it is assumed that individuals interact in loops such that a cooperative action can be returned, after several steps, to the original donor. According to Boyd and Richerson their model is unlikely to lead to a cooperative outcome, as it requires the loops to be relatively small, closed, and long-lasting. In Binmore's *Fun and Games'* (1992), the principle of indirect reciprocity is pithily resuméd as 'I won't scratch your back if you won't scratch their backs'. Binmore models this by imagining a world in which there are only two people alive at any stage, a mother and her daughter, with the daughter able to provide support to her mother.

The model considered by Nowak and Sigmund (1998a) is more in the tradition of evolutionary game theory. Consider a population of individuals having the options of helping another or not. In each generation, a number of potential donor-recipient pairs are chosen randomly: as before, this implies a cost c to the donor, if the help is actually provided, and a benefit b to the recipient. Furthermore, providing help increases the donor's score by one, whereas the score of a player refusing to help is decreased by one. (Note that the score has nothing to do with the payoff.) Initially all scores are zero. We consider strategies given by integers k; a player with such a strategy helps if and only if the score of the potential recipient is at least k. We can follow the frequencies of the strategies from generation to generation, allowing for occasional mutations.

A remarkably small number of interactions can lead to the emergence of cooperative populations where most members use $k = 0$ or $k = -1$ (for $b = 10$ and $c = 1$, an average of two interactions per lifetime suffices). If the simulation is continued, strategies which are less discriminating spread: players with $k = -5$, for instance, will rarely ever refuse to help, their score will therefore increase faster than average, and hence they will in turn be helped more often. But if the frequency of less discriminating players reaches a certain threshold, then defectors (players with $k = 5$, for instance, who practically never provide help) take over, so that cooperation disappears in the population. Once this happens, the average k-values will drop again, leading eventually back to a cooperative regime of players with maximal discrimination (*i.e.* $k = 0$).

To summarize, random drift can subvert populations of discriminate altruists by indiscriminate altruists; once their frequency is large, defectors can invade; but as soon as the defectors have reduced the proportion of indiscriminate altruists, the discriminate altruists can fight back and eliminate the defectors. This leads again to a cooperative population which is proof against defectors, but not against indiscriminate altruists, etc.

6. The good, the bad, and the discriminating

In order to obtain an analytic understanding, we can further simplify the model (see Nowak and Sigmund, 1998b), so that only two scores are possible, namely g (for 'good') and b (for 'bad'). Each player has two interactions per round, one as a donor and one as a recipient, against randomly chosen co-players. These two interactions are not with the same co-player. In fact, we may neglect the possibility that two players are ever paired twice. A player has score g if (and only if) he has provided help in the last round. Let us consider a population with three types of strategies only: type 1, the indiscriminate altruists (with frequency x_1); type 2, the defectors (with frequency x_2) and type 3, the discriminate altruists (with frequency $x_3 = 1 - x_1 - x_2$). Furthermore, we assume that in the first round, discriminators assume that the co-player has score g. It is easy to see that $P_i(1)$, the payoff for type i in the first round, is given by

$$P_1(1) = -c + b(x_1 + x_3)$$

$$P_2(1) = b(x_1 + x_3)$$

and

$$P_3(1) = -c + b(x_1 + x_3).$$

In the n-th round (with $n > 1$) it is

$$P_1(n) = -c + b(x_1 + x_3)$$

$$P_2(n) = bx_1$$

and

$$P_3(n) = (b - c)(x_1 + x_3^{n-1}x_2)/(x_1 + x_2).$$

If there is only one round per generation, then defectors win, obviously. This is no longer the case if there are N rounds, with $N > 1$. The total payoffs $\hat{P}_i := P_i(1) + ... + P_i(N)$ are given by

$$\hat{P}_1 = N[-c + b(x_1 + x_3)]$$

$$\hat{P}_2 = Nbx_1 + bx_3,$$

$$\hat{P}_3 = N(b - c) + x_2[-b + \frac{b - c}{1 - x_3}(1 + x_1 + ... + x_3^{N-1} - N)].$$

Let us now assume that the frequencies x_i of the three strategies evolve under the action of selection, with growth rates given by the difference between their payoff $\hat{P}_i$ and the average $\hat{P} = \sum x_i \hat{P}_i$. This yields the replicator equation

$$\dot{x}_i = x_i(\hat{P}_i - \hat{P})$$

on the unit simplex spanned by the three unit vectors $\mathbf{e}_i$ of the standard base. This equation has no fixed point with all $x_i > 0$, hence the three types cannot co-exist in the long run. The fixed points are the point $\mathbf{F}_{23}$ with $x_1 = 0$ and $x_3 + ... + x_3^{N-1} = c/(b - c)$, as well as all the points on the edge $\mathbf{e}_1\mathbf{e}_3$. Hence in the absence of defectors, all mixtures of discriminate and indiscriminate altruists are fixed points.

The overall dynamics can be most easily described in the case $N = 2$ (see Fig. 4). The parallel to the edge $\mathbf{e}_1\mathbf{e}_2$ through $\mathbf{F}_{23}$ is invariant. It consists of an orbit with ω-limit $\mathbf{F}_{23}$ and α-limit $\mathbf{F}_{13}$. This orbit l acts as a separatrix. All orbits on one side of l converge to $\mathbf{e}_2$. This means that if there are too few discriminating altruists, i.e. if $x_3 < c/(b - c)$, then defectors take over. On the other side of l,

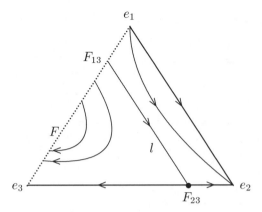

FIGURE 4. The population dynamics of discriminate and indiscriminate altruists and defectors (see text).

all orbits converge to the edge $\mathbf{e}_1\mathbf{e}_3$. The limit point lies somewhere between $\mathbf{e}_3$
and $\mathbf{F}$, the point with $x_3 = 2c/b$. In this case, the defectors are eliminated, and a
mixture of altruists gets established.

This leads to an interesting behaviour. Suppose that the society consists entirely of altruists. Depending on the frequency x_3 of discriminators, the state is
given by a point on the fixed point edge $\mathbf{e}_1\mathbf{e}_3$. We may expect that random drift
makes the state fluctuate along this edge and that from time to time, mutation
introduces a small quantity x_2 of defectors. What happens? If the state is between
$\mathbf{F}_{13}$ and $\mathbf{e}_1$, the defectors will take over. If the state is between $\mathbf{e}_3$ and $\mathbf{F}$, they will
immediately be selected against, and promptly vanish. But if a minority of defec-
tors invades while the state is between $\mathbf{F}$ and $\mathbf{F}_{13}$, something strange happens. At
first, the defectors thrive on the indiscriminate altruists and increase in frequency.
But thereby, they deplete their resource, the indiscriminate altruists. After some
time, the discriminate altruists take over and eliminate the defectors. The popu-
lation returns to the edge $\mathbf{e}_1\mathbf{e}_3$, but now somewhere between $\mathbf{e}_1$ and $\mathbf{F}$, where the
ratio of discriminate to indiscriminate altruists is so large that defectors can no
longer invade. The defectors have experienced a Pyrrhic victory. Their only hope
is that fluctuations will eventually decrease the frequency of discriminators again.
They have to wait until the state is between $\mathbf{F}_{13}$ and $\mathbf{e}_1$. For this, the fluctuations
have to cross the gap between $\mathbf{F}$ and $\mathbf{F}_{13}$. This takes some time. If defectors try
too often to invade, they will never succeed.

This behaviour corresponds precisely to the oscillations observed in the numerical simulations with the more sophisticated model (where the score k can take any
integer value). Long periods of cooperation are terminated when the frequency of
indiscriminate altruists becomes too large; this is immediately followed by a sharp
increase in the number of defectors. Next, these defectors are eliminated by the
discriminators, and then another long period of cooperation begins.

7. Know your partner

So far we have assumed that all players know each other's score. This is not
realistic, of course. Even in small groups, where all members know each other
well, there must be many interactions that are not watched by everyone. We must

assume that players can have different scores in the eyes of different co-players. If we consider numerical simulations where an interaction between two individuals is observed by a random subset of the other individuals, and assume that only these 'on-lookers' have the possibility to update their perception of the donor's image score, the information is contained in a matrix whose elements s_{ij} denote the image score of player i as seen by player j. In a donor-recipient interaction between j and i, player j will cooperate if $s_{ij} \geq k_j$. If j has no information on i then $s_{ij} = 0$.

The model now depends on the probability that a given individual observes an interaction between two other individuals. We find again that cooperation can easily be established and dominate the population, but a larger number of interactions per generation is needed. For larger groups, it is more difficult to establish cooperation, because the fraction of individuals that obtain information about any particular interaction will be smaller. Therefore, more interactions are required (relative to group size) in order to discriminate against defectors.

We can investigate this analytically by extending the previous two-score model. Let us assume that in each round, a player can be donor or recipient with the same probability $1/2$, and that q is the probability that a given individual knows the score of a randomly chosen opponent. A discriminator who does not know the score of the co-player will assume with probability 1 that this score is g. If g_n denotes the frequency of g-scorers in the population in round n, and $x_{1g}(n), x_{2g}(n)$ and $x_{3g}(n)$ are the frequencies of g-scoring indiscriminate altruists, unconditional defectors resp. discriminators in round n, then clearly $x_{1g}(n) = x_1$ and $x_{2g}(n) = (1/2)x_{2g}(n-1)$, since a defector is with probability $1/2$ in the role of a donor and then unmasks himself. Therefore

$$x_{2g}(n) = \frac{x_2}{2^{n-1}} .$$

The score of a discriminator remains unchanged if he is a recipient. If he is a potential donor, he will either know the co-player (with probability q) and help if the co-player has score g (as happens with probability g_{n-1}), or else he will not know the co-players score, and help. Since this latter alternative holds with probability $1 - q$, this yields

$$x_{3g}(n) = (1/2)x_{3g}(n-1) + (1/2)x_3(1 - q + qg_{n-1}).$$

Since $g_n = x_{1g}(n) + x_{2g}(n) + x_{3g}(n)$, it follows that

$$g_n = sg_{n-1} + \frac{1}{2}(x_1 + (1-q)x_3)$$

with $s = (1 + qx_3)/2$. This recurrence relation implies (together with $g_1 = 1$) that

$$g_n = (\frac{1 + qx_3}{2})^{n-1} \frac{x_2}{1 - qx_3} + \frac{x_1 + (1-q)x_3}{1 - qx_3} .$$

The payoff for the indiscriminate altruists in round n is

$$P_1(n) = -(c/2) + (b/2)(x_1 + x_3).$$

The payoff $P_2(n)$ for the unconditional defectors depends on their score. Those with score b receive $b(x_1 + (1-q)x_3)/2$ and those with score g in addition $qbx_3/2$, so that

$$P_2(n) = (b/2)[x_1 + (1-q)x_3 + x_3q(x_{2g}(n)/x_2)] .$$

Finally, a discriminator receives $[-c(qg_n + 1 - q) + bx_1 + (1 - q)bx_3)]/2$ if he has score b, and in addition $bqx_3/2$ if he has score g, so that we obtain

$$P_3(n) = -(c/2)(qg_n + 1 - q) + (b/2)(x_1 + x_3) - (b/2)qx_3[1 - (x_{3g}(n)/x_3)].$$

Instead of assuming a fixed number of rounds per generation, let us suppose that it is a random variable. If we assume for instance that $w < 1$ is the probability for a further round, then the total payoff for type i is given by $\hat{P}_i = \sum P_i w^n$. A straightforward computation shows that the phase portrait of the corresponding replicator equation $\dot{x}_i = x_i(\hat{P}_i - P)$ looks as before. The separatix l is now given by $x_3 = c(2 - w)/bwq$. In one region, all orbits converge to $\mathbf{e}_2$, *i.e.* defectors take over. In the other region, all orbits converge to a point on the edge $\mathbf{e}_1\mathbf{e}_3$ which depends on the initial value. In particular, altruism can only get established if

$$q > c/b$$

which looks exactly like Hamilton's rule, except that the coefficient of relatedness is replaced by what we may call a coefficient of acquaintanceship, *i.e.* the probability that a player knows the score of the co-player. If this condition holds, then the average number of rounds per generation, *i.e.* $1/(1-w)$, must exceed $(bq+c)/(bq-c)$. It is only in this case that in a population consisting of defectors and discriminators, the discriminating strategy is stable.

8. Discussion

It should also be mentioned that this discriminator strategy is related to, but different from the T_1-strategy proposed by Sugden (1986). In Sugden's model, in each round a randomly chosen player needs help, and each of the other players can provide some help (thus the needy player can get as payoff $(m-1)b$, where m is the group size). Sugden's T_1- strategy is based on the concept of *good standing*, just as in the case of direct reciprocation. Players are born with a good standing, and keep it long as they help all needy players with good standing. If not, they lose their good standing. Sugden argues that such a strategy can work as a basis for an insurance principle within the population. We stress that a player can keep his good standing by refusing to help someone of bad standing, whereas in our model, he would lose his g-score whenever he refuses help, even if the potential recipient is a b-scorer. Sugden's T_1 strategy is more sophisticated, but like cTFT, it is vulnerable to errors in perception (see Boerlijst *et al.*, 1997). If information is incomplete, then a player observed while withholding his help may be misunderstood; he may have defected on a player with good standing, or punished someone with bad standing. An eventual error can spread. The discriminator rule is less demanding on the player's mental capabilities.

Moralistic aggression, social norms and feelings of fairness and solidarity have evolved in hominid groups in order to manage the complex social relations within their families, bands and tribes (Ridley, 1996). Human decisions and activities are not only guided by economic considerations, but to a large extent by a complex household of emotions. These emotions and passions are themselves the outcome of biological necessities, in particular the need for division of labor and for cooperation. The way of life of the first hominids entering the savannah did not allow for autarchy, but required a highly adapted social life based on reciprocation and an internalized system of norms – a natural law in the fullest sense.

The importance of trading acts of mutual assistance has been recognized for a very long time – we may find it, for instance, in David Hume's book on *Human Nature* (1740): 'I learn to do service to another, without bearing him any real kindness: because I foresee that he will return my service, in expectation of another of the same kind. I foresee that he will return my service, in expectation of another of the same kind, and in order to maintain the same correspondence of good offices with me *and others* [italics added]. And accordingly, after I have served him and he is in possession of the advantage arising from my action, he is induced to perform his part, as foreseeing the consequences of his refusal.'

An even older formulation of the principle of indirect reciprocation can be found in the biblical injunction 'Give and to you it shall be given'.

It may well be that indirect reciprocation works in other species. A hopeful candidate is the Arabian babbler. Zahavi (1995) mentions that individuals often compete with each other, jostling for the role of a donor. They interfere with the helping of others in feeding the nestlings, in allofeeding between adults, in sentinel activities, in mobbing and in the defense of the common territory. As Zahavi notes, this is difficult to interpret in terms of kin selection, group selection or (direct) reciprocation. If the beneficiary gets help from someone else, so much for the better, it would seem. But if helping is a means for raising one's score, competition for being the one who gives help makes perfect sense.

We cannot expect that direct and indirect reciprocation are always well separated. Alexander (1987) views indirect reciprocity 'as a consequence of direct reciprocity occurring in the presence of interested audiences – groups of individuals who continually evaluate the members of their society as possible future interactants'. The smooth transition from direct to indirect reciprocity is neatly captured in Pollock and Dugatkin (1992), who studied direct reciprocation in the usual context of the repeated Prisoner's Dilemma and allowed the players to occasionally observe a co-player before starting the repeated interaction. They analyzed a strategy called Observer Tit For Tat (oTFT). If the future co-player was seen defecting in his last interaction, then oTFT prescribes to defect in the first round. Pollock and Dugatkin were mostly interesting in comparing this strategy with the usual TFT, and showed that it is not always advantageous. This seems surprising at a first glance – why should additional information be a handicap? The reason is that the conclusion drawn by oTFT is not always appropriate. A player last seen defecting can be a defector, but could also be a TFT player punishing a defector, in which case the new interaction starts with the wrong move. But Pollock and Dugatkin also remarked that oTFT may be better able to invade a resident population of defectors (*viz.* with a smaller initial cluster), and that it can be stable against defectors when TFT is not. They also stressed that the probability of future encounters among specific players engaged in an iterated Prisoner's Dilemma game may in principle decline to zero, *i.e.* that oTFT can hold its own against defectors when no degree of future interaction with the current partner was presumed. In this case, oTFT reduces precisely to the discriminator strategy. This shows that a gradual transition from direct to indirect reciprocation can lead to the establishment of a discriminating strategy based image scores.

The simple thought experiments performed by means of automata show that self-assembly can lead to norms enforcing cooperation, both through internalization of fair and tolerant rules by means of inner states and through social pressure based on one's standing within the community.

References

[1] Alexander, R.D. (1979), *Darwinism and Human Affairs*, Univ. Washington Press, Seattle, WA.

[2] Alexander, R.D. (1987) *The Biology of Moral Systems*, Aldine de Gruyter, New York.

[3] Axelrod, R. (1984) *The Evolution of Cooperation*, reprinted 1989 by Penguin, Harmondsworth.

[4] Axelrod, R. (1997), *The Complexity of Evolution*, Princeton Studies in Complexity, Princeton UP.

[5] Axelrod, R. and W.D. Hamilton (1981), The evolution of cooperation, *Science* **211**, 1390-6.

[6] Binmore, K.G. (1992), *Fun and Games: a Text on Game Theory*, Heath and Co, Lexington, Massachussetts.

[7] Boerlijst, M., M.A. Nowak and K. Sigmund (1997), The Logic of Contrition, *Journ. Theor. Biol.*, **185**, 281-293.

[8] Boyd, R. and P.J. Richerson (1989), The evolution of indirect reciprocity, *Social Networks* **11**, 213-36.

[9] Dawkins, R. (1976) *The Selfish Gene*, Oxford UP, second ed. 1989.

[10] Hamilton, W.D. (1963), The evolution of altruistic behaviour, *American Naturalist* **97**, 354-6.

[11] Hofbauer, J. and K. Sigmund (1998), *Evolutionary Games and Population Dynamics*, Cambridge UP.

[12] Hume, D. (1740), *A Treatise of Human Nature*, reprinted in Penguin Classics 1985, Penguin, Harmondsworth.

[13] John Paul II (1997), Message to the Pontifical Academy of Science, *Quart. Rev. Biol.* **72**, 381-3.

[14] Leimar, O. (1997), Repeated games: a state space approach, *Journ. Theor. Biol.*, **184**, 471-98.

[15] Lindgren, K. (1991) Evolutionary phenomena in simple dynamics. In *Artificial life II* (ed. C.G. Langton *et al.*), Santa Fe Institute for Studies in the Sciences of Complexity, Vol. X, 295-312.

[16] May, R.M. (1987), More evolution of cooperation, *Nature* **327**, 15-17.

[17] Maynard Smith, J. (1982), *The Theory of Games and Evolution*, Cambridge UP.

[18] Maynard Smith, J. and E. Szathmáry (1995), *The Major Transitions in Evolution*, Freeman, Oxford.

[19] Nowak, M.A. and K. Sigmund (1993), Win-stay, lose-shift outperforms tit-for-tat, *Nature*, **364**, 56-8.

[20] Nowak, M.A. and K. Sigmund (1998a), Evolution of indirect reciprocity by image scoring, to appear in *Nature*.

[21] Nowak, M.A. and K. Sigmund (1998b), The dynamics of indirect reciprocity, submitted to *Journ. Theor. Biol.*

[22] Pollock, G.B. and L.A. Dugatkin (1992), Reciprocity and the evolution of reputation, *Journ. Theor. Biol.* **159** 25-37.

[23] Ridley, M. (1996), *The Origins of Virtue*, Penguin, Harmondsworth.

[24] Selten, R. (1975), Re-examination of the perfectness concept for equilibrium points in extensive games, *International Journal of Game Theory* **4**, 25-55.

[25] Sigmund, K. (1995), *Games of Life*, Penguin, Harmondsworth.

[26] Sugden, R. (1986) *The Evolution of Rights, Co-operation and Welfare*. Blackwell, Oxford.

[27] Trivers, R. (1971), The evolution of reciprocal altruism, *Quarterly Review of Biology* **46**, 35-57.

[28] Trivers, R. (1985), *Social Evolution*, Menlo Park. CA, Benjamin Cummings.

[29] Weibull, J. (1995) *Evolutionary Game Theory*, MIT Press, Cambridge, MA.

[30] Wilson, D.S. and E. Sober (1994), Re-introducing group selection to human behavioural sciences, *Behavioural and Brain Sciences* **17**, 585-654.

[31] Zahavi, A. (1995), Altruism as a handicap – the limitations of kin selection and reciprocity, *Journ. of Avian Biology* **26**, 1-3.

E-mail address: Karl.Sigmund@esi.ac.at

INSTITUT FÜR MATHEMATIK, UNIVERSITÄT WIEN, STRUDLHOFGASSE 4, A-1090 VIENNA, AND IIASA, LAXENBURG, AUSTRIA

Lectures on Mathematics in the Life Sciences
Volume **26**, 1999

The Linear Geometry of Genetic Operators
with Applications to the Analysis of Genetic Drift
and Genetic Algorithms using Tournament Selection

Lothar M. Schmitt and Chrystopher L. Nehaniv

ABSTRACT. We extend the results of the authors in [*Theo. Comp. Sci.*, **200** (1998) 101-134] on the convergence of genetic algorithms with proportional fitness-scaled selection to genetic algorithms with various types of tournament fitness selection often used in practical applications. Our analysis allows for a general notion of fitness selection which includes types of fitness selection mechanisms where the fitness of the individual depends upon the population it lives in (such as frequency-dependent fitness selection). We also include a short analysis of the effect of genetic drift which has applications to the study of genetic algorithms with selection and crossover but without mutation. Genetic algorithms are represented by Markov chains on probability distributions over the set of all possible populations of a fixed finite size. Linear analysis of the stochastic matrices representing the three phases mutation, crossover, and fitness selection of a genetic algorithm — using, *e.g.*, computation of spectra and Hilbert space techniques — yields new insight into the geometric properties of these phases. We show by explicit estimates, that for small mutation rates a genetic algorithm asymptotically spends most of its time in uniform populations regardless of the crossover rate. We show ergodicity of genetic algorithms using scaled multiple-bit mutation, and a wide range of fitness selection mechanisms which includes scaled proportional fitness selection, tournament fitness selection, and fitness selection mechanisms where the fitness of the individual depends upon the population it lives in. Our analysis permits varying mutation and crossover rates according to a pre-determined schedule, where the mutation rate stays bounded away from zero. If proportional fitness selection, tournament fitness selection or rank selection is used, then the limit probability distribution of such a process is fully positive at all populations of uniform fitness regardless of the fitness selection method and its possible scaling.

1. Introduction

Genetic algorithms with proportional fitness selection have previously been investigated by many authors (see, *e.g.*, [**DP1, Gol1, GS1, Rdo1, SNF1, Suz1, Vos1**]). The purpose of the present paper is to extend the results of [**SNF1**] on

1991 *Mathematics Subject Classification.* Primary: 92D15, 47N60, 47N10 and Secondary: 92B05, 92D25, 15A51, 68U99.

Both authors are main contributors to this paper.

the convergence of genetic algorithms with proportional fitness-scaled selection to genetic algorithms with various types of tournament fitness selection [**Gol2, Mic1, Mit1**] and other types of selection often used in practical applications. As an application of our model for genetic algorithms, a simple mathematical analysis of the effect of genetic drift (*cf.* [**MS1, Rou1**]) is also included in this paper. It has applications to the study of genetic algorithms with selection and crossover but without mutation.

Our model for genetic algorithms is a Markov chain model. Previously, such models for genetic algorithms using proportional fitness selection without scaling have been developed, *e.g.*, in [**DP1, GS1, Rdo1, Vos1**], but mostly on a different underlying vector space. In our[1] model, the basis vectors of the underlying vector space V correspond to the possible populations in the genetic algorithm which are represented as bitstrings[2]. The state of the algorithm after t steps or generations is described by a probability distribution $v = \sum \lambda_p p$, $(0 \leq \lambda_p \leq 1, \sum \lambda_p = 1)$, over all possible populations $p \in \wp$. For each of the genetic operators, mutation $M_{\mu(t)}$ with mutation rate $\mu(t)$, crossover $C_{\chi(t)}$ with crossover rate $\chi(t)$, and fitness selection F_t, we have a corresponding stochastic operator on the underlying vector space V. A single iteration G_t through one step t of a genetic algorithm is given by the product linear operator $G_t = F_t C_{\chi(t)}^K M_{\mu(t)}$. If $M_{\mu(t)}$, $C_{\chi(t)}$, and F_t are constant, then the genetic algorithm is called *simple* and is described by a single linear operator $G = G_t$.

Our analysis starts by investigating each of the operators $M_{\mu(t)}$, $C_{\chi(t)}$, and F_t separately using methods of linear algebra. This is subsequently combined to an overall analysis of the associated genetic algorithm.

In section 3, we shall include an analysis of multiple-bit mutation, the most commonly used method of mutation in genetic algorithms. We shall obtain estimates for the coefficient of ergodicity of G_t based upon estimates for the coefficient of ergodicity of $M_{\mu(t)}$ providing a complete proof of [**Suz1**, p. 60: Lemma 1]. In addition, we shall investigate in Proposition 3.2 the interplay of mutation and the fitness selection operator in regard to the change in non-uniformity over the course of the algorithm.

In section 4, first we shall consider what we call *simple crossover*, which is probably used most in applications. Then we shall discuss convergent behavior of iterated crossover operations on certain subspaces of V called Hardy-Weinberg spaces, which are the span over populations with (locally) constant allele frequencies. One important aspect investigated in Proposition 4.4 (see also [**SNF1**, Prop. 10] for a related result) is the understanding of how the crossover operation enhances the action of mutation in the random search phase $C_{\chi(t)}^K M_{\mu(t)}$ ($K \in \mathbf{N}$ fixed) of the algorithm.

In section 5, we discuss a variety of fitness selection schemes used in applications of genetic algorithms. But our analysis also treats types of fitness selection

---

[1] The vector space representation developed in [**SNF1**] was found independently by Rudolph [**Rdo1**], who mainly analyses the convergence of simple genetic algorithms including a strategy to keep track of a best-so-far creature seen by the algorithm (a strategy related to but distinct from the so-called 'elitist strategy').

[2] Many authors (see, *e.g.*, [**DP1, Vos1**]) model populations as unordered multi-sets. These models for genetic algorithms can be obtained as projections of our model simply by forgetting the order of creatures within populations. Our model frees the initial description from clumsy combinatorial coefficients, which may hinder a subsequent detailed analysis.

mechanisms where the fitness of the individual depends upon the ambient population, including for example cases in which the fitness of a genotype depends on the frequencies of the various types of individuals in the population (*cf.*, [**AH1, Sig1**]). In particular, we analyze how fitness selection schemes contract towards the space of probability distributions over uniform populations. This is later combined with the analysis of the interplay fitness selection vs. mutation of Proposition 3.2 mentioned above to obtain estimates for the proportion over non-uniform populations of the limit probability distribution for a possibly scaled genetic algorithm. These results hold with a very general notion of crossover which is formulated in section 4. Our presentation illuminates the discussion of "punctuated equilibrium" in [**Vos1**]. [**Vos1**] contends that a simple genetic algorithm is near local optima most of the time, visiting every state infinitely often.

It is well-known (*e.g.*, [**MS1, Rou1, CK1**]) from the analysis of finite populations on which mutation does not act and which are of uniform fitness (equivalently: there is no selective pressure), that these change in the course of evolution to uniform populations containing solely multiple copies of one particular genotype. We investigate the effect of genetic drift with selection and crossover but without mutation. It is not hard to show that such an algorithm converges (pointwise on distributions over populations) to a distribution over uniform populations. We give an analysis establishing such convergence and discussing average time to total genetic uniformity in a setting allowing quite general types for selection and recombination (crossover). Classical drift is a special case, but we allow also nontrivial selection including not only proportional fitness selection and tournament fitness selection but also types of fitness selection mechanisms where the fitness of the individual depends upon the population in which it finds itself. This analysis extends findings in [**Fog1**], and yields insight into the success of some applications of genetic algorithms using crossover without mutation.

In section 7, we show ergodicity of a large variety of genetic algorithms. These algorithms can be scaled in regard to the mutation rates for multiple-bit mutation, and in regard to the crossover rates for a very general setting for crossover. We impose as main condition, that the mutation rates $\mu(t)$ for multiple-bit mutation converge to a strictly positive value. The algorithm can use tournament fitness selection, rank selection or scaled fitness selection but also fitness selection mechanisms where the fitness of the individual depends also on the population as discussed in section 5. Our analysis applies to most standard methods of fitness scaling such as linear fitness scaling [**Gol1**, p. 77], sigma-truncation [**Gol1**, p. 124], and power law scaling (see [**Gol1**, p. 124], and section 5.2). The behavior of fitness-scaled genetic algorithms in regard to convergence was posted as an open question in [**Rdo1**]. This question can be answered in the negative for convergence to an optimal individual, a population containing an optimal individual, or even to a distribution strictly positive only over such populations, and in the positive in the sense of convergence to a unique probability distribution over all populations. We show that the limit probability distribution of such processes is fully positive at populations of uniform fitness. For tournament fitness selection and a large set of fitness scaling methods, *i.e.*, those with so-called strong fitness scaling such as power law scaling, the limit distribution depends only on the pre-order or rank induced by the fitness function on the set of creatures, and not the particular values of the fitness function, *i.e.*, in the limit, difference but not degree of difference matters. This is in agreement with the viewpoint of Baker [**Bak1**] as discussed in section 5.4, who proposed to use only the fitness ranking of individuals in determining fitness selection probabilities.

Many of our results in this paper complement or generalize results in [**SNF1**]. We have included here some additional material from [**SNF1**], in order to have a somewhat self-contained, "single-track" presentation.

2. Notation

The number s of creatures in a population is constant through time, and a creature in a population is given as a string of bits[3] of length ℓ. Consequently, a population can be seen as a bitstring of length $L = s \cdot \ell$. Let $N = 2^L$. Let V be the free vector space over the set $\wp = \{p_0, \ldots, p_{N-1}\}$ of so defined populations, *i.e.*, the basis of V is labeled by bitstrings of length L, or the corresponding binary integers. This induces a canonical order on the basis of V. We shall identify a population $p \in \wp$ with the associated basis vector in V.

We shall consider the standard inner product in V. If W is a subspace of V spanned by a subset $\wp_1$ of $\wp$, then e_W is, by definition, the vector in W such that $<e_W, p> = 1/dim(W)$, for every $p \in \wp_1$. In particular, we set $e = e_V$. Let P_e denote the orthogonal projection onto the subspace generated by e. For any vector $v = \sum \alpha_p p$ in V, the ℓ^1-*norm* or *Hamming norm* of v is given by $||v||_1 = \sum |\alpha_p|$, while $||v||_2 = (\sum |\alpha_p|^2)^{1/2}$ is the ℓ^2-*norm* or *Euclidean norm*. If $X : V \to V$ is a linear map, then the corresponding operator norms [**Sha1**, p. 5: (5)] will also be denoted by $||X||_1$ resp. $||X||_2$.

Let $S \subset V$ be the set of probability distributions over populations, *i.e.*, S is the positive $|| \cdot ||_1$-unit-sphere in V.

Let $U \subset V$ be the vector space spanned by the set of uniform populations, *i.e.*, spanned by the set of all populations that contain s copies of one particular creature. Let P_U denote the orthogonal projection onto U.

We let matrices operate by multiplication from the left. The *spectrum* of a matrix X will be denoted by $sp(X)$. A matrix or vector is called *fully positive*, if all its entries are strictly positive. A matrix X is called *[column] stochastic*, if its entries are non-negative reals and its columns all sum to 1.

If X is stochastic, then let the *coefficient of ergodicity with respect to the ℓ^1-norm*, be given by $\tau_r(X) = \max\{||Xv||_1 : v \in \mathbf{R}^N \subset V, v \perp e, ||v||_1 = 1\}$ (see [**Sen1**, p. 138: Lemma 4.2]). A useful fact from [**Sen1**, p. 137], who uses matrix multiplication from the right and row stochastic matrices, gives:

(2.1) $$1 - \tau_1(X) \geq \sum_{q \in \wp} \min_{p \in \wp} <q, Xp>.$$

---

[3]Some biologists may object that the genotype (in this case a bit-string) is identified with the 'creature', since organisms in nature are *not* uniquely determined by their genome, but by an interplay of genetic, developmental, physical, and environmental factors. While we certainly agree, such usage is justified when in fact that fitness is a function of genotype as in most optimization genetic algorithms. Other rigorous analyses of genetic algorithms that we know of make this assumption of a well-defined genotype-to-fitness map. As a step towards more biological realism, our analysis goes beyond this: in addition to static or scaled fitness functions of genotypes, we also treat types of fitness selection in which the probability of an individual being selected may depend not only on its genotype but also on the those of the rest of the population (frequency-dependent selection). For various examples of such frequency-dependent selection see [**Sig1**].

3. Mutation

3.1. The matrix describing multiple-bit mutation. Most implementations of genetic algorithms use multiple-bit mutation defined as follows: an application of the mutation operator consists of independently deciding for each bit of a population, whether or not to change it with probability (or *mutation rate*) $\mu \in (0, 1/2]$. The matrix M_μ associated with multiple-bit mutation is doubly stochastic, and is given by [**SNF1**, Prop. 3.1]:[4]

$$(3.1) \qquad <q, M_\mu p> \;=\; \mu^{\Delta(p,q)} (1-\mu)^{L-\Delta(p,q)}, \quad p, q \in \wp,$$

where $\Delta(p, q)$ is the number of bit positions in which p differs from q.

It is immediate from equation 3.1 and inequality 2.1 that $\tau_1(M_\mu) \leq 1 - (2\mu)^L$. If we consider crossover operations and fitness selection operations[5] that can be described by stochastic matrices, then the above in combination with, *e.g.*, [**Sen1**, p. 137: Thm. 4.8] or [**IM1**, p. 151: Thm. V.3.2], yields a complete proof of [**Suz1**, p. 60: Lemma 1]: *If the mutation rates $\mu(t)$, $t \in \mathbf{N}^+$, satisfy $\mu(t) \geq 2^{-1} t^{-1/L}$, then the inhomogeneous Markov chain $G_t \cdot G_{t-1} \cdots G_1$ is weakly ergodic, where $G_t = F_t C^K_{\chi(t)} M_{\mu(t)}$ as in the introduction.*

Next, let us give an recursive tensor product construction for M_μ, and use this to determine its spectrum. For this purpose, simply look at populations as bitstrings of length L. Let $M_\mu^{(L)}$ denote the stochastic matrix describing multiple-bit mutation on bitstrings of length L. We have:

$$(3.2) \qquad M_\mu^{(1)} \;=\; \begin{pmatrix} 1-\mu & \mu \\ \mu & 1-\mu \end{pmatrix}.$$

A straightforward induction argument shows that $M_\mu^{(L+1)} = M_\mu^{(L)} \otimes M_\mu^{(1)}$. This allows to compute the spectrum of $sp(M_\mu)$ using $sp(M_\mu^{(L+1)}) = sp(M_\mu^{(L)}) \cdot sp(M_\mu^{(1)})$. In fact, $sp(M_\mu) = \{(1-2\mu)^i : 0 \leq i \leq L\}$ [**SNF1**, Prop. 3.4].

Observe that by Perron's Theorem [**Sha1**, p. 23: Cor. 1], e is up to scalar multiples the unique eigenvector to eigenvalue 1 of the fully positive matrix M_μ. Thus, the components $(\mathbf{1} - P_e)v$ of $v \in S$, which are orthogonal to e, are contracted by M_μ towards 0 geometrically with

$$\|M_\mu(\mathbf{1} - P_e)v\|_2 \;\leq\; (1-2\mu)\,\|(\mathbf{1} - P_e)v\|_2.$$

By the above, $1 - 2\mu$ is the 2nd largest eigenvalue of M_μ, and the latter estimate is easily obtained by representing M_μ as diagonal matrix over an orthonormal basis of V consisting of eigenvectors of M_μ. Thus, the high-dimensional simplex S, which is the convex hull of the unit vectors, *i.e.*, populations $p \in \wp$, is contracted geometrically by mutation towards its barycenter e. Figure 1 shows the overall situation for the degenerate case of populations that consist of exactly two one-bit creatures. On the other hand, fitness selection (see the definition of generalized fitness selection in section 5.1 and the remarks closing that section) contracts towards the space of uniform populations which is shown in Figure 1 left.

[**SNF1**, Sec. 2.1] contains a more detailed discussion of multiple-bit mutation as well as so-called *single-bit mutation* where only one bit is randomly flipped in a

---

[4]Note that $<q, Xp>$ is just the probability that applying a stochastic operator X to population p results in population q.

[5]*E.g.* fitness selection given by a genotype-to-fitness map combined with one of the selection mechanisms described in section 5

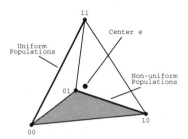

FIGURE 1. Simplex of probability distributions over populations
on which genetic operators act: probability distribution e, with all
populations equally likely, is the uniquely determined fixed point
of mutation. Thick lines indicate probability distributions over
uniform populations (left) and non-uniform populations (right).

single mutation step. As mentioned above, multiple-bit mutation is used most in
implementations of genetic algorithms. Single-bit mutation is in better agreement
with the philosophy of "small neighborhoods" used in simulated annealing [**AL1**].
Both mutation operators contract probability distributions geometrically towards
e. By examining the spectra of both mutation operators, one finds that this contraction stays controlled in the sense that for sufficiently small μ the Euclidean
length of a vector perpendicular to e cannot shrink too much [**SNF1**, Prop. 2.4
and Prop. 4.4].

3.2. On the interplay between mutation and fitness selection. Suppose
that $T(t)$, $t \in \mathbf{N}$, is a sequence of stochastic matrices, which leave the linear space
U over uniform populations invariant, and otherwise contract towards U. Let us
discuss an algorithm that can be modeled as $G_t = T(t) M_\mu$, $\mu \in [0, \frac{1}{2}]$ fixed. We
are interested in estimates that describe how products of the G_t shrink the non-
uniform portion of a particular $v \in S$. Later, we shall set $T(t) = F_t C_{\chi(t)}^K$ where
F_t is fitness selection, and $C_{\chi(t)}$ is the crossover operator in a genetic algorithm.
Proposition 3.1 is listed next for the convenience of the reader, and is contained in
[**SNF1**, Prop. 4.4]. Proposition 3.2 generalizes part of [**SNF1**, Lemma 13].

PROPOSITION 3.1. *If $v \in S$ is a probability distribution over populations and
$h = \lfloor \frac{s}{2} \rfloor$ where s is the population size, then we have:*

$$\|(\mathbf{1} - P_U) M_\mu v\|_1 \leq \quad \left(1 - (\mu^s + (1-\mu)^s)^\ell\right) \|P_U v\|_1 +$$
$$+ \left(1 - ((1-\mu)^h \mu^{s-h} + \mu^h (1-\mu)^{s-h})^\ell\right) \|(\mathbf{1} - P_U) v\|_1.$$

PROOF. (Sketch) Using a combinatorial argument, the first term in the sum to
the right is obtained by explicitly determining $(\mu^s + (1-\mu)^s)^\ell$ as the probability of
generating uniform populations from a fixed uniform population. The second term
is determined by showing that $((1-\mu)^h \mu^{s-h} + \mu^h (1-\mu)^{s-h})^\ell$ is a lower estimate
for the probability of generating uniform populations from a fixed non-uniform
population. $\quad\Box$

PROPOSITION 3.2. *Let $T(t)$, $t \in \mathbf{N}$, be a sequence of stochastic matrices, and
$\theta \in [0, 1)$ be such that $T(t) P_U v = P_U v$, and $\|(\mathbf{1} - P_U) T(t) v\|_1 \leq \theta \|(\mathbf{1} - P_U) v\|_1$ for
every $v \in S$. Let $\mu \in [0, \frac{1}{2}]$ be fixed, and $G_t = T(t) M_\mu$. Let $h = \lfloor \frac{s}{2} \rfloor$.*

1. We have: $\beta = (\mu^s + (1-\mu)^s)^\ell - ((1-\mu)^h \mu^{s-h} + \mu^h (1-\mu)^{s-h})^\ell \in [0,1]$.
2. We have: $\|(\mathbf{1} - P_U) G_t v\|_1 \leq \theta \, (1 - (\mu^s + (1-\mu)^s)^\ell + \beta \|(\mathbf{1} - P_U) v\|_1)$.
3. In addition, we have for $t \in \mathbf{N}$:

$$\|(\mathbf{1} - P_U) \prod_{k=t}^{1} G_k v\|_1 \leq \theta \frac{1 - (\mu^s + (1-\mu)^s)^\ell}{1 - \theta\beta} + \theta^t \beta^t \|(\mathbf{1} - P_U) v\|_1.$$

PROOF. By examining the function $k \mapsto (1-\mu)^k \mu^{s-k} + \mu^k (1-\mu)^{s-k}$, one obtains that β lies in $[0,1]$. The $\|\cdot\|_1$-inequality for $T(t)$ and Proposition 3.1 imply:

$$\begin{aligned}\|(\mathbf{1} - P_U) G_t v\|_1 &\leq \theta \, \|(\mathbf{1} - P_U) M_\mu v\|_1 \\ &\leq \theta \, (1 - (\mu^s + (1-\mu)^s)^\ell + \beta \|(\mathbf{1} - P_U) v\|_1).\end{aligned}$$

The inequality in Proposition 3.2(3) is obtained by applying inequality 3.2(2) inductively to $\|(\mathbf{1} - P_U) \prod_{k=t}^{1} G_k v\|_1$, and using a geometric series estimate. $\square$

4. Crossover

The crossover operation models the effect of genetic recombination in populations. In this section, we shall mainly give a listing of properties of so-called simple crossover, which illustrate how crossover assists the mutation operation in the random search phase $C_{\chi(t)}^K M_{\mu(t)}$ ($K \in \mathbf{N}$ fixed) of the genetic algorithm. For additional detail and proofs, the reader is referred to [**SNF1**, Section 2.2]. For some convergence theorems, which we shall prove in later sections of this paper, only a few general properties of the crossover operation are needed, which we shall establish first.

4.1. Generalized crossover. An *generalized crossover operation* is a continuous map $\chi \mapsto C_\chi$, $\chi \in [0,1]$, into the set of self-adjoint stochastic matrices such that $C_\chi q = q$ for every uniform population $q \in \wp \cap U$, and $\chi \in [0,1]$.

LEMMA 4.1. *If C_χ, $\chi \in [0,1]$, is a generalized crossover operation, then:*
1. $C_\chi P_U = P_U = P_U^* = P_U C_\chi$, *and* $C_\chi (\mathbf{1} - P_U) = C_\chi - P_U = (\mathbf{1} - P_U) C_\chi$.
2. $\chi \mapsto C_\chi^K$, $K \in \mathbf{N}$ *fixed, is also a generalized crossover operation.*

4.2. Simple crossover. Simple crossover as defined below is perhaps the most widely used crossover operation in applications of genetic algorithms. Simple crossover can be described in terms of so-called *elementary crossover operations* $C_{i,j,r}$ which are defined next.

Let a population $p = (c_1^{old}, \ldots, c_s^{old})$ consist of creatures c_i^{old}, each of which is a binary string of length ℓ. For indices $i < j \in \{1, \ldots, s\}$, and a *crossover point* $r \in \{1, \ldots, \ell\}$, we define an elementary crossover operation $C_{i,j,r}$ on p as the following procedure: If $c_i^{old} = b_1^{(i)} \ldots b_\ell^{(i)}$, and $c_j^{old} = b_1^{(j)} \ldots b_\ell^{(j)}$ as bit-strings, then replace p by an identical population, except that in the ith and in the jth position one has the creatures

$$c_i^{new} = b_1^{(j)} \ldots b_r^{(j)} b_{r+1}^{(i)} \ldots b_\ell^{(i)}, \quad \text{and} \quad c_j^{new} = b_1^{(i)} \ldots b_r^{(i)} b_{r+1}^{(j)} \ldots b_\ell^{(j)}.$$

Note that if $r = \ell$, this amounts to exchanging c_i and c_j as complete creatures[6]. Intuitively, an elementary crossover is the exchange of genetic material, *i.e.*, the

---

[6]The case $k = \ell$ is included for mathematical convenience. It could be excluded, and the analysis that follows could be carried out with modified definitions of the Hardy-Weinberg spaces defined below. In fact, by excluding the case $k = \ell$, one has to keep track of the last bits of

alleles at the first r loci, between two genomes of the population. An elementary crossover operation $C_{i,j,r}$ induces a self-adjoint, unitary, stochastic matrix[7] on V, which we shall also denote by $C_{i,j,r}$. One has $C_{i,j,r}^2 = \mathbf{1}$, and $sp(C_{i,j,r}) = \{-1, 1\}$.

For the discussion of simple crossover, assume that the size s of populations $p \in \wp$ is even. Let $R = (r_1, \ldots, r_{s/2})$ be a vector of $s/2$ crossover points, and fix a *crossover rate* $\chi \in [0, 1]$. Now, we define the crossover operation $C_{R,\chi}$ with crossover points R and crossover rate χ to be the operator

$$C_{R,\chi} = \prod_{i=1}^{s/2} ((1 - \chi)\mathbf{1} + \chi C_{2i-1,2i,r_i}).$$

This means that an application of $C_{R,\chi}$ to a population p consists of $s/2$ independent operations to the sequentially paired off creatures of p. In each of these $s/2$ operations, either the current population is left unchanged with probability $1 - \chi$, or an elementary crossover operation $C_{2i-1,2i,r_i}$ is applied with probability χ to manipulate $(c_{2i-1}^{old}, c_{2i}^{old})$. Since $sp(A\,B) \subset sp(A) \cdot sp(B)$ for commuting matrices A and B, and all possible $C_{2i-1,2i,r_i}$ mutually commute, we have

$$sp(C_{R,\chi}) \subset \{(1 - 2\chi)^i : 0 \le i \le s/2\}.$$

Finally, to capture the usual simple crossover operator (see for example [**Gol1**, p. 64], [**Mit1**, p. 8], or [**Hol1**, p. 97]) in our model, we must randomize the choice of crossover points so that all points have equal probability. The probability for a particular crossover point vector R to be selected randomly is $\ell^{-s/2}$. Thus, we obtain the self-adjoint stochastic *simple crossover operator* C_χ by summing over all choices for R:

$$C_\chi = \ell^{-s/2} \sum_R C_{R,\chi}.$$

C_χ commutes with mutation since all elementary crossover operations $C_{i,j,r}$ commute with mutation. In addition, we can give an estimate for the second largest eigenvalue of C_χ since $sp(A+B) \subset sp(A) + sp(B)$ for commuting matrices A and B, which has an obvious generalization to convex combinations of commuting matrices (here: the $C_{R,\chi}$). In fact, we have

$$(4.1) \qquad sp(C_\chi) \subset [-1 + \delta, 1 - \delta] \cup \{1\}, \quad \text{where} \quad \delta = \ell^{-s/2}(1 - |1 - 2\chi|).$$

The spectral estimate just obtained can be used to show how simple crossover assists the mutation operation M_μ as an averaging procedure, *i.e.*, how simple crossover accelerates convergence towards e, if powers of $C_\chi^K M_\mu$ ($K \in \mathbf{N}$ fixed) are applied to a fixed $v \in S$. For this purpose, we need some notation. Let $p = (c_1, c_2, \ldots, c_s) \in \wp$, all c_i being bitstrings of length ℓ. Then we set:

$$Mean(p) = (\tfrac{1}{2}(c_1 + c_2), \ldots, \tfrac{1}{2}(c_{s-1} + c_s)) \in \{0, \tfrac{1}{2}, 1\}^{\ell \cdot \frac{1}{2} s}.$$

One can think of $Mean(p)$ as a list of $s/2$ gene frequency vector for local pairs in p. For a given $\xi \in \mathbf{R}^{\ell \cdot \frac{1}{2} s}$, let V_ξ, the *Hardy-Weinberg*[8] *space of populations with gene*

---

creatures in a population in the definition of $Mean(p)$ given below since a crossover operation could never switch the parents' c_i^{old} and c_j^{old} last bits.

[7]$C_{i,j,r}$ is up to a rearrangement of the basis $\wp$ of V a direct sum of the identity matrix on a subspace of V and flip matrices $M_1^{(1)}$ (see formula 3.2).

[8]In the case of randomly mating (diploid) populations in population genetics, Hardy and Weinberg observed that allele frequencies within the population remain unchanged from generation to generation [**MS1**, pp. 31-36]. The classical Hardy-Weinberg result states that, for infinite

frequency ξ, be defined as

$$V_\xi = span_\mathbf{C}\{p : p \in \wp, Mean(p) = \xi\}.$$

In addition, let $D = span_\mathbf{C}\{e_{V_\xi} : \xi = Mean(p), \text{ for some } p \in \wp\}$. Let P_ξ, repectively P_D, be the orthogonal projections onto V_ξ, repectively D. Observe that $P_e P_D = P_e = (P_e P_D)^* = P_D P_e$ since $e \in D$. Consequently,

$$(4.2) \quad \mathbf{1} = (P_e + (\mathbf{1} - P_e))P_D + (\mathbf{1} - P_D) = P_e + (\mathbf{1} - P_e)P_D + (\mathbf{1} - P_D).$$

LEMMA 4.2. *Let M_μ be the doubly stochastic matrix describing multiple-bit mutation. Then we have $M_\mu D \subset D$. Consequently, $M_\mu P_D = P_D M_\mu P_D = (P_D M_\mu P_D)^* = P_D M_\mu$, and $M_\mu (\mathbf{1} - P_D) = (\mathbf{1} - P_D) M_\mu$.*

PROOF. Consider for a moment the stochastic matrix B that models flipping the first bit in a population $p \in \wp$ with probability μ and the $(\ell + 1)^{\text{th}}$ bit with probability μ, and leaving all other bits unchanged. Let $p_{00} \in \wp$, and $\xi = (\xi_1, \ldots, \xi_{s/2}) = Mean(p_{00})$ such that $\xi_1 = 0$. This implies that both bit 1 and bit $\ell + 1$ in p_{00} are 0. Let $p_{ij} \in \wp$, $i, j \in \{0, 1\}$, be equal to p_{00} except that the first bit in p_{ij} equals i and the $(\ell + 1)^{\text{th}}$ bit equals j. Then we have:

$$B\,p_{00} = (1 - \mu)^2 p_{00} + \mu(1 - \mu)(p_{01} + p_{10}) + \mu^2 p_{11}.$$

The latter expression is, in particular, symmetric in regard to $Mean(p_{01})$. Hence, $B\,e_{V_\xi} \subset D$. A similar discussion can be carried out for $\xi_1 = 1/2$ and $\xi_1 = 1$. Finally, observe that M_μ is a product of matrices "B" for $L/2$ pairs of bits.[9] $\qquad\square$

LEMMA 4.3. *With the notation defined prior to Lemma 4.2, we have:*

1. *Each V_ξ is an invariant subspace for all $C_{2i-1,2i,r_i}$, $C_{R,\chi}$, and for C_χ.*
2. *C_χ decomposes into a block diagonal matrix with blocks C_χ^ξ acting on V_ξ.*
3. *C_χ leaves D invariant. Consequently, $C_\chi P_D = P_D = P_D^* = P_D C_\chi$, and $C_\chi (\mathbf{1} - P_D) = C_\chi - P_D = (\mathbf{1} - P_D) C_\chi$.*
4. *If $\chi \in (0, 1)$, and C_χ^ξ is seen as a matrix acting on V_ξ, then e_{V_ξ} is up to scalar multiples the only eigenvector of C_χ^ξ to eigenvalue 1.*

PROOF. (1), (2), and (3) are obvious from the definitions. Let $p \in \wp$, and $\xi = Mean(p)$. If $q \in \wp$ such that $\xi = Mean(q)$, then one can apply a series of elementary crossover operations $C_{2i-1,2i,r_i}$ "from left to right" to the bitstring of p to generate q. Hence, $<(C_\chi^\xi)^n p, q> \neq 0$ for a suitable $n \in \mathbf{N}$. Thus, C_χ^ξ is irreducible by [**Sha1**, Prop. I.6.2]. Since by hypothesis $\chi \in (0, 1)$ assures a positive diagonal of C_χ^ξ, (4) follows from Perron-Frobenius theory [**Sha1**, p. 23: Cor. 2]. $\qquad\square$

The next result illuminates the mechanism by which the crossover operation assists mutation in the random search phase $C_\chi^K M_\mu$ ($K \in \mathbf{N}$ fixed) of the genetic algorithm. If this phase is iterated without fitness evaluation, then $(C_\chi^K M_\mu)^n p$, $n \in \mathbf{N}$, starting with initial population $p \in \wp$ probabilistically explores the whole search space making the chance for every population to be generated small but equal. In particular, populations with "good" creatures have an equal chance to be generated. If $C_\chi^K M_\mu$ is used as the proposal scheme for a simulated annealing type algorithm as outlined in [**SNF1**, Remark on Simulated Annealing], then the

---

populations in which selection and mutation are not involved, allele frequencies remain constant in randomly mating diploid populations. The analogous invariance here for our haploid, finite populations under crossover makes this name appropriate.

[9][**SNF1**, Prop. 3] contains Lemma 4.2, but the proof given here is much simpler.

individual steps of the algorithm can be seen as a stochastic sequence of operators $C_\chi^K M_\mu$ and $\mathbf{1}$. In particular in this situation, the "spreading out of the search space" is heuristically illustrated by the next result which complements [**SNF1**, Prop. 10].

PROPOSITION 4.4. *Let C_χ as above describe simple crossover, and let M_μ describe multiple-bit mutation*[10]. *Let δ be as in equation 4.1. If $v \in S$, then $v = e + d(v) + o(v)$, where $d(v) = (\mathbf{1} - P_e) P_D v$, and $o(v) = (\mathbf{1} - P_D) v$. Set $w = C_\chi^K M_\mu v$ with fixed $K \in \mathbf{N}$. Then we have*

$$d(w) = M_\mu d(v), \qquad o(w) = M_\mu C_\chi^K o(v),$$
$$\|d(w)\|_2 \le (1 - 2\mu) \|d(v)\|_2, \quad and \quad \|o(w)\|_2 \le (1 - 2\mu)(1 - \delta)^K \|o(v)\|_2.$$

PROOF. First, observe that $<v - e, e> = 0$. This shows that $P_e v = e$. Thus, $v = e + d(v) + o(v)$ by equation 4.2. By Lemma 4.2, and Lemma 4.3 we have: $d(w) = (\mathbf{1} - P_e) P_D C_\chi^K M_\mu v = (\mathbf{1} - P_e) P_D M_\mu v = M_\mu (\mathbf{1} - P_e) P_D v = M_\mu d(v)$, and similarly $o(w) = o(v)$, since $1 - P_D$ commutes with C_χ and M_μ. Now, the $\|\cdot\|_2$-estimates follow from the facts that: (1) e, resp. the e_{V_ξ}, are up to scalar multiples the unique fixed points for M_μ, resp. the C_χ^ξ. (2) We have computed the spectra of M_μ, respectively C_χ, which determine $\|\cdot\|_2$-contracting factors for components orthogonal to the eigenvectors to eigenvalue 1.   □

4.3. Random Mating. In [**SNF1**, Sec. 2.2], both the above simple crossover and a so-called *unrestricted crossover*, which does not restrict elementary crossover to adjacent pairs but randomly selects mated pairs, are investigated. For the latter, results similar to the above hold (see [**SNF1**, Prop. 9, Prop. 10]). It is interesting to note, that unrestricted crossover is linked to a commuting family of representations of the group of permutations of s elements. This can effectively be used to compute spectral estimates [**SNF1**, Prop. 9.7]. Note that for the crossover operations discussed so far, iteration of the crossover operator induces crossover operations with multiple crossover points.

Similarly to mutation, both crossover operations exhibit convergent behavior, but convergence occurs on the Hardy-Weinberg spaces V_ξ of constant allele frequency. See [**SNF1**, Sec. 1.3] for the definition of the Hardy-Weinberg spaces V_ξ for unrestricted crossover. Both crossover operations contract geometrically towards a family of fixpoints e_{V_ξ}, and the vector space V spanned by populations $p \in \wp$ decomposes into blocks containing exactly one such fixpoint. Thus, there is strong analogy between the action of crossover and that of mutation: Mutation mixes globally by tending to make all populations equally likely. Crossover mixes locally by tending toward making all populations in each fixed Hardy-Weinberg subspace equally likely.

5. Fitness Selection

Fitness selection in genetic algorithms is modeled on reproductive success in nature. In many applications, the selection pressure is modeled in the following way: given a finite collection C of creatures in a model "world", a *fitness function* $f : \mathsf{C} \to \mathbf{R}^+$ is defined which determines the chances of survival for $c \in \mathsf{C}$ in every step of the genetic algorithm. In our setting C is the set of bitstrings of length ℓ.

---

[10]The reader may establish an analogous result for single-bit mutation (*cf.* [**SNF1**, Sec. 2.1]).

A genetic algorithm becomes a function optimizer, if the task is then to find an element $c \in C$ such that $f(c)$ is maximal.

Implementations of genetic algorithms mostly use the following fitness selection strategies: proportional fitness selection, tournament fitness selection, and rank selection. Before we discuss these types of fitness selection strategies in detail, let us summarize all in a definition which includes also types of fitness selection where the fitness of the individual depends upon the population it lives in. In particular, frequency-dependent fitness is included in this notion of generalized fitness selection (see Examples 2 and 3 below).

5.1. Generalized fitness selection. An *generalized fitness selection scaling* is a map $t \mapsto F_t$, $t \in \mathbf{N}$, into the set of stochastic matrices, such that

1. $F_t q = q$ for every uniform population $q \in \wp \cap U$, and $t \in \mathbf{N}$.
2. There exist disjoint subsets $\wp_1$ and $\wp_2$ of $\wp$ such that $\wp \cap U \subset \wp_1$, and $\wp_1 \cup \wp_2 = \wp$. Let $W_\kappa = span_{\mathbf{C}}(\wp_\kappa)$, $\kappa = 1, 2$.
3. We have for some fixed $\tau \in \mathbf{N}$, fixed $\theta \in [0, 1)$, and every $v \in S$, $t \in \mathbf{N}$:
   $$P_{W_1} F_t P_{W_1} v = F_t P_{W_1} v, \quad P_{W_1} \left(\prod_{k=\tau+t}^{t} F_k \right) P_{W_2} v = \left(\prod_{k=\tau+t}^{t} F_k \right) P_{W_2} v,$$
   and $\|(\mathbf{1} - P_U) F_t P_{W_1} v\|_1 \leq \theta \|(\mathbf{1} - P_U) P_{W_1} v\|_1$.

EXAMPLE 1: Suppose the fitness function $f : C \to \mathbf{R}^+$ is injective.[11] If $p \in \wp \setminus U$ is non-uniform, then eliminate one of the identical creatures in p with lowest fitness value, and replace it by a creature with highest fitness value. As final step, permute the creatures in the population locally or globally. In this setting, $\wp_1$ is the set of populations which are uniform or uniform up to one element of lower fitness, and $\theta = 0$. $\wp_2$ is the set of all other populations, and $\tau = s - 3$.

A generalized fitness selection scaling need not necessarily be defined in terms of a fitness function $f : C \to \mathbf{R}^+$. For example, we can allow that there is a fitness function $f(c, p, t)$ depending upon creature $c \in C$, population $p \in \wp$ and $t \in \mathbf{N}$. We include 2 examples of this type:

EXAMPLE 2: Consider a cyclic relation denoted $\lhd$ defined on C as follows: for $c, d \in C$, we have $c \lhd d$, if $0 \leq c \leq d \leq 2^\ell - 1$ as binary integers and $d - c \leq 2^{\ell-1}$, or if $0 \leq d \leq c \leq 2^\ell - 1$ as binary integers and $c - d \geq 2^{\ell-1}$. Thus, $c \lhd d$, if it is shorter or the same distance to ascend cyclically from c to d, than from d to c. Let $p \in \wp$ contain c. We define the fitness $f(c, p) > 0$ of c relative to p as the probability that c satisfies $c \lhd d$ with a randomly chosen element $d \in p$. We can now scale $f(c, p)$ to define $f(c, p, t)$ by setting

$$f(c, p, t) = f(c, p)^{g(t)}$$

where $g(t) \geq 1$, and $\lim_{t \to \infty} g(t) = \infty$. We can then perform proportional fitness selection with $f(c, p, t)$ as described below in section 5.2. In this situation, we have $\wp = \wp_1$ and $\theta = 1 - s^{-s}$ (see the discussion preceding Lemma 5.3).

EXAMPLE 3: We can also scale a fitness function $f : C \to \mathbf{R}^+$ locally. Let g be as in Example 2, and set

$$f(c, p, t) = f(c)^{g(t \cdot \nu(p))}$$

where $\nu(p)$ is the number of distinct creatures in p. This procedure scales the selection pressure such that in populations which have fewer distinct creatures the

---

[11][**SNF1**, Sec. 2.3] discusses how to replace a non-injective fitness function f by an injective fitness function such that the objective of the genetic algorithm to find a creature in C of maximal fitness for f is retained.

chances of better creatures are not so much increased as in populations which are very diverse. In this situation, we have also $\wp = \wp_1$, and $\theta = 1 - s^{-s}$.

The last two examples show that our approach can model what are called "frequency dependent" types of selection. However, in what follows we shall concentrate on fitness selection mechanisms that are commonly used in implementations of genetic algorithms on computers.

LEMMA 5.1. *Let $t \mapsto F_t$, $t \in \mathbf{N}$, be a generalized fitness selection scaling. Then we have: $F_t P_U = P_U$, and $(1 - P_U) F_t = (1 - P_U) F_t (1 - P_U)$.*

LEMMA 5.2. *Let $t \mapsto F_t$, $t \in \mathbf{N}$, be a generalized fitness selection scaling with associated $\theta \in [0, 1)$ and $\tau \in \mathbf{N}$. Then,*

$$\left\| (1 - P_U) \prod_{k=t}^{1} F_k \, v \right\|_1 \leq \theta^{t-\tau-1} \left\| (1 - P_U) v \right\|_1$$

for every $v \in S$, and every $t \in \mathbf{N}$ such that $t > \tau + 1$.

PROOF. Using $P_{W_1} F_k P_{W_1} v = F_k P_{W_1} v$, we have by the $\|\cdot\|_1$-inequality for F_t:

$$\left\| (1 - P_U) \left(\prod_{k=t}^{\tau+2} F_k \right) P_{W_1} v \right\|_1 \leq \theta^{t-\tau-1} \left\| (1 - P_U) P_{W_1} v \right\|_1.$$

for $t \in \mathbf{N}$. Using the second identity in Lemma 5.1, we have for $t > \tau + 1$:

$$\left\| (1 - P_U) \left(\prod_{k=t}^{\tau+2} F_k \right) \left(\prod_{k=\tau+1}^{1} F_k \right) v \right\|_1 \leq \theta^{t-\tau-1} \left\| (1 - P_U) P_{W_1} \prod_{k=\tau+1}^{1} F_k \, v \right\|_1 =$$

$$= \theta^{t-\tau-1} \left\| (1 - P_U) F_{\tau+1} (1 - P_U) \prod_{k=\tau}^{1} F_k \, v \right\|_1 \leq \cdots \leq \theta^{t-\tau-1} \left\| (1 - P_U) v \right\|_1.$$

since $1 - P_U$ is a diagonal matrix, and $\|F_k\|_1 = 1$ (see [**Sha1**, p. 5: (5)]). $\square$

Thus the linear analysis of fitness selection yields the following geometric picture: The operators F_t act as the identity operators on U, the span over uniform populations (shown in Figure 1 left), and shrink the space spanned over non-uniform populations at a geometric rate determined by θ.

5.2. Proportional fitness selection. Proportional fitness selection has probably been investigated the most as part of genetic algorithms. See, *e.g.*, [**DP1, GS1, Rdo1, Suz1, Vos1**]. Since we have discussed a very general notion of fitness selection in [**SNF1**, Sec. 2.3] which includes proportional fitness selection but also *linear fitness scaling* [**Gol1**, p. 77], and *sigma-truncation* [**Gol1**, p. 124], we shall list here only a few definitions.

At the population level, the fitness operator F_t, $t \in \mathbf{N}$, yields a new population from the current population, based on a fitness-proportional probability for each creature to be reproduced in the next generation: Let $f_t(c)$ be the (possibly) scaled fitness value[12] of creature $c \in \mathsf{C}$ at step t of the genetic algorithm. A common choice for a fitness scaling used with proportional fitness selection is so-called *power law scaling* [**Gol1**, p. 124], [**Suz1**, p. 65] where $f_t(c_i) = (f(c_i))^{g(t)}$, and $\lim_{t \to \infty} g(t) =$

---

[12]We shall assume that a scaling **never** changes the rank of creatures in a population, *i.e.*, the order on populations induced by the fitness function f. Hence for every $t \in \mathbf{N}$, $f(c_1) < f(c_2)$ is equivalent to $f_t(c_1) < f_t(c_2)$, for all $c_1, c_2 \in \mathsf{C}$. The fitness $f_t(c)$ here depends only on the creature $c \in \mathsf{C}$ and possibly on time (step, generation) $t \in \mathbf{N}$ deterministically.

∞.[13] For a population $p = (c_1, c_2, \ldots, c_\sigma) \in \wp$ consisting of creatures $c_i \in \mathsf{C}$, $1 \leq i \leq s$, the probability that any given new creature in step $t+1$ is produced by literally copying c_i is $f_t(c_i)/\sum_{j=1}^{s} f_t(c_j)$.

It is not hard to determine the associated stochastic matrix F_t [**SNF1**, Prop. 11.1]: for $p = (c_1, c_2, \ldots, c_s) \in \wp$, $c_1, \ldots, c_\sigma \in \mathsf{C}$, and $q = (d_1, d_2, \ldots, d_s) \in \wp$, $d_1, \ldots, d_\sigma \in \mathsf{C}$, let $n(d_i, p)$ denote the number of copies of d_i in the population p. The probability that q is generated from p by fitness selection is given by

$$(5.1) \qquad <q, F_t p> \;=\; \prod_{i=1}^{s} \frac{n(d_i, p)\, f_t(d_i)}{\sum_{j=1}^{s} f_t(c_j)}.$$

In case power law scaling with $\lim_{t\to\infty} g(t) = \infty$ is used, then $F_\infty = \lim_{t\to\infty} F_t$ exists. F_∞ only depends upon the rank in the population and not the actual fitness values and is the same for all so-called *strong fitness scalings* (see [**SNF1**, pp. 123-124]). As outlined in section 5.4, Baker [**Bak1**] proposed to use only the fitness ranking of individuals for fitness selection. The above result on F_∞ is used in [**SNF1**, Thm. 18] and Theorem 7.4 to show that asymptotically, Baker's criterion (rank) is the only essential factor in the limit probability distribution of a genetic algorithm using power law scaling with $\lim_{t\to\infty} g(t) = \infty$.

For a creature c_{max} of maximal fitness in a population p, the chance that the next population consist entirely of copies of c_{max} is at least $s^{-s} > 0$. From this, it is immediate that scaled proportional fitness selection is a generalized fitness selection with $\theta = 1 - s^{-s}$, $\wp_1 = \wp$, and $\wp_2 = \emptyset$. Similarly, one obtains:

LEMMA 5.3. *Let $p = (c_1, c_2, \ldots, c_s) \in \wp$, $c_1, \ldots, c_\sigma \in \mathsf{C}$, be of uniform fitness, i.e., $f_t(c_1) = f_t(c_2) = \ldots = f_t(c_s)$. In this situation, $<p, F_t p> \geq s^{-s}$.*

5.3. Tournament fitness selection. Mitchell [**Mit1**, p. 170] proposes the following tournament selection mechanism: Fix $\phi \in [0, \frac{1}{2})$. For $j = 1, \ldots, s$, do: STEP 1: Select two creatures c_{i_1} and c_{i_2} ($1 \leq i_1, i_2 \leq s$) at random from the current population $p = (c_1, \ldots, c_s) \in \wp$. It is assumed that all creatures $c_1, \ldots, c_s$ participate in this random selection process in both rounds, *i.e.*, a particular c_i may be selected twice. STEP 2: If the selected creatures c_{i_1} and c_{i_2} have the same fitness, then select one of them with probability $\frac{1}{2}$. Otherwise, set d_j to the creature with lower fitness value with probability ϕ, and to the creature with higher fitness value with probability $1 - \phi$.

Let F_ϕ be the stochastic matrix acting on V associated with the tournament selection mechanism described above. Let $p \in \wp$, $c_{max} \in \mathsf{C}$ be a ceature of maximal fitness in p, and $q = (c_{max}, \ldots, c_{max}) \in \wp \cap U$. It is easy to see that

$$(5.2) \qquad <q, F_\phi p> \;\geq\; s^{-s} = 1 - \theta_\phi$$

Goldberg [**Gol2**] proposes to exclude a chosen element from the second random round in the determination of the tournament pair, and to set $\phi = 0$, *i.e.*, the better creature is always chosen. As a variant of the latter, Michalewicz [**Mic1**, p. 59] proposes to chose one of the best out of a randomly selected tuple of creatures of size σ ($2 \leq \sigma \leq s - 1$) from the old population p, and similar to the above repeat this procedure s times to fill out a new population q. Observe that in these cases, the number of distinct creatures in the new population q is strictly less than the

---

[13]A variant of such power law scaling is so-called *Boltzmann selection* [**Mit1**, pp. 168–169] which is inspired by the simulated annealing algorithm [**AL1**].

number of distinct creatures in the old population p, if p is of non-uniform fitness and the fitness function f is injective.

All the tournament fitness selection mechanisms discussed in this section satisfy the requirements of a generalized fitness selection scaling as defined in section 5.1, if one sets F_t to an appropriate constant operator, $\wp_1 = \wp$, and $\wp_2 = \emptyset$. Determining the associated contracting factors θ other than θ_ϕ listed in equation 5.2 is left to the reader.

Observe that the tournament fitness selection mechanisms discussed in this section only depend upon the rank of creatures, and not the actual fitness values.

5.4. Rank selection. Baker [**Bak1**] proposed to use only the fitness ranking of individuals in determining fitness selection probabilities, *i.e.*, in this approach, the fitness function only determines a ranking in the current population. Creatures are selected for the next step of the genetic algorithm, *e.g.*, by scaled or unscaled proportional fitness selection where the unscaled probability of survival of a creature $c \in \mathsf{C}$ is proportional to the rank of c in the current population. See [**Bak1**] and [**Mit1**, p. 170] for versions of genetic algorithms whose fitness selection schemes use rank.

5.5. Introducing spatial structure. At this point, we can naturally introduce spatial structure into our model, since one may identify position in the population with a location in a world of any topology a researcher would like to model. For example, one could investigate genetic algorithms with local variation of the mutation rate, the crossover rate, or the fitness function modeling different local conditions for mutation, mating, reproduction and survival such as radiation exposure, abundance of resources and environmental factors. The study of evolution of spatially structured populations is an area of active interest for both computer scientists [**Mit1**] and evolutionary ecologists and population geneticists [**Rou1, PYW1**]. This represents a substantial advantage of our model over the multiset-based models of, *e.g.*, [**DP1, Vos1**] which do not capture any spatial structure. Moreover, these models may be obtained as simple projections of ours, so there is much to gain and nothing to lose in keeping track of order of creatures in the population.

6. Genetic Drift: Convergence under Fitness Selection and Crossover

Genetic drift, well-known from treatises on evolutionary genetics (see for example [**MS1**, pp. 24-27],[**CK1**],[**Rou1**]), refers to the phenomenon that populations not subject to mutation tend to become uniform, *i.e.*, over the generations an increasing number of creatures (individuals) in the population become genetically identical. In this section, we investigate genetic drift for a genetic algorithm with fitness selection but without mutation. It is not hard to show that such an algorithm converges (pointwise on distributions over populations) to distribution over uniform populations.

Consider a generalized crossover operation $\chi \mapsto C_\chi$, $\chi \in [0, 1]$, and a generalized fitness selection scaling F_t, $t \in \mathbf{N}$ as defined in section 4.1, and 5.1, respectively. Set $H_t = \prod_{k=t}^{1} F_k C_{\chi(k)}$.

We claim that $\lim_{t \to \infty} P_U H_t v_0$ and consequently $\lim_{t \to \infty} \| (\mathbf{1} - P_U) H_t v_0 \|_1$ always exist. We have $F_k C_{\chi(k)} P_U = P_U$, and $P_U^* q = P_U q = q$ for $q \in \wp \cap U$. Hence,

$$<H_t v_0, q> \; = \; <H_{t-1} v_0, q> + <F_t C_{\chi(t)} (\mathbf{1} - P_U) H_{t-1} v_0, q>, \quad q \in \wp \cap U.$$

Thus, the q-component of $H_t v_0$ is increasing (and bounded) for every uniform population q as $t \to \infty$ (which is intuitively rather obvious).

Let us first consider the case of generalized fitness selection where $\wp_1 = \wp$, and $\wp_2 = \emptyset$ in the definition in section 5.1. Proposition 3.2(3) for $\mu = 0$, and $T(k) = F_k C_{\chi(k)}$ yields $\lim_{t \to \infty} \|(\mathbf{1} - P_U) H_t v_0\|_1 = 0$, since $\theta < 1$.

Now consider the case with $\wp_2 \neq \emptyset$. Suppose that $\chi_0 \in [0, 1]$, and $\alpha \in (0, 1]$, exist such that $C_\chi = \alpha \mathbf{1} + C_\chi^o$ for $\chi \in [0, \chi_0]$, where C_χ^o is a matrix with positive entries. In the case of simple crossover and of unrestricted crossover as in [**SNF1**, Sec. 2.2], this condition can be satisfied for every $\chi_0 \in [0, 1)$. Suppose, that $\chi(t) \in [0, \chi_0]$ for every $t \in \mathbf{N}$. Since $\|C_\chi\|_1 = 1$, we have $\|C_\chi^o\|_1 = 1 - \alpha$ for $\chi(t) \in [0, \chi_0]$ by [**Sha1**, p. 5: (7')].

We have by Lemma 5.1 for $n \in \mathbf{N}$ such that $n \leq t$:

$$\|(\mathbf{1} - P_U) H_t v_0\|_1 \leq \alpha \|(\mathbf{1} - P_U) F_t H_{t-1} v_0\|_1 + (1 - \alpha) \|(\mathbf{1} - P_U) H_{t-1} v_0\|_1 \leq$$

$$\leq \alpha^n \left\|(\mathbf{1} - P_U) \left(\prod_{k=t}^{t-n+1} F_k \right) H_{t-n} v_0 \right\|_1 + (1 - \alpha^n) \|(\mathbf{1} - P_U) H_{t-n} v_0\|_1.$$

Hence, for $n = \tau + 2$ we have by Lemma 5.2, and the fact that $\theta < 1$:

$$\lim_{t \to \infty} \|(\mathbf{1} - P_U) H_t v_0\|_1 \leq (\alpha^{(\tau+2)} \theta + (1 - \alpha^{(\tau+2)})) \lim_{t \to \infty} \|(\mathbf{1} - P_U) H_{t-\tau-2} v_0\|_1 = 0.$$

Both cases considered show the phenomenon of genetic drift in a very general setting: we impose no requirement on convergence of the crossover operators or fitness selection operators themselves. If one wants to remove the condition $\chi(t) \in [0, \chi_0]$ on the crossover rates $\chi(t)$ in the second case, then one has to analyze the interplay of the crossover operators $C_{\chi(t)}$ with the sets $\wp_1$ and $\wp_2$.

The average time for convergence to a uniform population can be computed as

$$\left\| P_U \sum_{k=0}^{\infty} k \left(\prod_{t=k}^{1} (F_t C_{\chi(t)} (\mathbf{1} - P_U)) \right) e \right\|_1 < \infty.$$

This expression depends upon the specific interplay between the crossover operation, and fitness selection, and a detailed evaluation is beyond the scope of this article. In the case that $\wp_2 = \emptyset$, an upper bound is given by $(1 - 2^{\ell-L})/(1 - \theta)^2$. A similar bound can be derived in the case $\wp_2 \neq \emptyset$ using Lemma 5.1 and the additional assumptions on crossover as above. This is left to the reader.

Some researchers have used genetic algorithms without mutation with good results. The above shows that the convergence of these experiments to uniform populations can be seen from the linear geometry of the algorithm. The fact that uniform populations with optimal creatures are favored in such an algorithm can be understood heuristically from the averaging feature of crossover on the Hardy-Weinberg spaces. By the proof of Lemma 4.3(4), the sub-blocks of C_χ^ξ of the crossover matrix C_χ are irreducible with uniquely determined fixpoint e_{V_ξ}. Thus, on each Hardy-Weinberg V_ξ, C_χ acts as a mixing matrix which tends to make every population in V_ξ equally likely. Fitness selection may result in changing Hardy-Weinberg spaces in the course of the algorithm. However, a modest fitness selection pressure will likely allow the algorithm to stay in Hardy-Weinberg spaces of large dimension for a long period of time. In that case, the overall mixing effect of the crossover operation (as a substitute for the mixing effect of mutation) makes appearance of optimal creatures likely. These creatures are then favored by the fitness operation.

7. Ergodicity for Genetic Algorithms

In the final section of this exposition, we shall establish strong ergodicity for inhomogeneous Markov chains associated with scaled genetic algorithms. Our analysis applies to genetic algorithms using tournament fitness selection and many other fitness selection methods such as linear fitness scaling [**Gol1**, p. 77], sigma-truncation [**Gol1**, p. 124], and power law scaling as discussed in section 5.2. Our result also applies to types of fitness selection mechanisms where the fitness of the individual also depends on the ambient population (see Examples 2 and 3 in section 5.1). In the proof of our main theorem 7.4, the results of the next subsection are used which deal with convergence of inhomogeneous Markov chains.

7.1. Mathematical results on inhomogeneous Markov chains.

If G is any stochastic matrix, then *the canonical positive projection* $P_1(G)$ onto the space of invariant vectors of G is given by $P_1(G) = \lim_{k \to \infty} \frac{1}{k} \sum_{\kappa=0}^{k-1} G^\kappa$ (*cf.* [**Sha1**, Prop. I.3.4]). $P_1(G)$ is a stochastic matrix as limit of convex combinations of such.

LEMMA 7.1. *Let G be a fully positive, stochastic matrix acting on V with smallest entry bounded below by $\gamma > 0$. Let, as before, $N = dim(V)$.*

1. *G has a one-dimensional eigenspace to eigenvalue 1, which is generated by a strictly positive $v \in S$.*
2. *Every eigenvalue $\lambda \neq 1$ of G satisfies $|\lambda| \leq 1 - N\gamma$.*
3. *If Γ is any circle around 1 with radius less than $N\gamma$, then the canonical positive projection $P_1(G)$ onto the one-dimensional eigenspace to eigenvalue 1 is given by*

$$P_1(G) \;=\; \frac{1}{2\pi i} \int_\Gamma (\zeta \mathbf{1} - G)^{-1}\, d\zeta \;=\; (v, \ldots, v).$$

PROOF. (1) By Perron's Theorem [**Sha1**, p. 23: Cor. 1], 1 is a simple root of the characteristic equation of G. Hence, the eigenspace of G to eigenvalue 1 is one-dimensional. [**Sha1**, Prop. I.6.2(c)] shows that G possesses a strictly positive eigenvector $v \in S$ to eigenvalue 1. (2) is [**Sha1**, p. 40: Cor.]. To show (3), we use that $G = X J X^{-1}$, where X is an invertible matrix, and J is a Jordan matrix for G. We have by [**Sha1**, Prop. I.3.4] and Cauchy's formula:

$$P_1(G) \;=\; \lim_{\zeta \to 1} (\zeta - 1)\,(\zeta \mathbf{1} - G)^{-1} \;=\; \lim_{\zeta \to 1} X\,(\zeta - 1)\,(\zeta \mathbf{1} - J)^{-1}\, X^{-1} \;=\;$$

$$=\; X \left(\frac{1}{2\pi i} \int_\Gamma (\zeta \mathbf{1} - J)^{-1}\, d\zeta \right) X^{-1} \;=\; \frac{1}{2\pi i} \int_\Gamma (\zeta \mathbf{1} - G)^{-1}\, d\zeta.$$

Finally observe that for every $p \in \wp$, the vector $P_1(G)\,p$ must be a multiple of v, and must be a probability distribution since $P_1(G)$ is stochastic. $\square$

LEMMA 7.2. *Let $G_t = M_{\mu(t)}\, F_t\, C_{\chi(t)}$ for some fixed $t \in \mathbf{N}$. G_t is a fully positive, stochastic matrix with smallest entry bounded below by $\mu(t)^L$. Consequently, we have for the spectrum of G_t:* $|sp(G_t)| \subset [0, 1 - (2\mu(t))^L] \cup \{1\}$.

PROOF. Each entry of $M_{\mu(t)}\, F_t\, C_{\chi(t)}$ is a convex combination of the entries in the rows of $M_{\mu(t)}$, which are all bounded below by $\mu(t)^L$. The statement about $sp(G_t)$, follows from Lemma 7.1(2). $\square$

Lemma 7.1(3) allows us to give a simplified proof of [**SNF1**, Thm. 16] which equals [**Gid1**, Thm. 1.1]. However, [**Gid1**] only proves a continuous-time analogue of [**Gid1**, Thm. 1.1].

THEOREM 7.3. *Let G_t, $t \in \mathbf{N}$, be a sequence of fully positive, stochastic matrices such that $G_\infty = \lim_{t \to \infty} G_t$ exists and is fully positive. In accordance with Lemma 7.1(1), let v_t, v_∞, resp., w_t be the fully positive probability distributions, which are up to scalar multiples the uniquely determined fixed points of G_t, G_∞, resp., $H(t) = \prod_{k=t}^{1} G_k$. In this situation, we have*

$$P_1(G_\infty) = \lim_{t \to \infty} H(t), \quad and \quad v_\infty = \lim_{t \to \infty} v_t = \lim_{t \to \infty} w_t.$$

PROOF. Since $G_\infty = \lim_{t \to \infty} G_t$, we conclude by the discussion of coefficients of ergodicity $\tau_1(\cdot)$ in section 2 of this paper and Lemma 7.1(2), that there exists an $\epsilon > 0$ such that $\tau_1(G_t) \leq 1 - \epsilon$ for all $t \in \mathbf{N} \cup \{\infty\}$, and the second largest eigenvalue of G_t is less than $1 - \epsilon$ for all $t \in \mathbf{N} \cup \{\infty\}$. Let Γ be the circle in the complex plane around 1 of radius $\epsilon/2$. By Lemma 7.1(3), we have

$$P_1(G_t) = \frac{1}{2\pi i} \int_\Gamma (\zeta\mathbf{1} - G_t)^{-1} \, d\zeta.$$

This expression is continuous in G_t as long as the modulus of the second largest eigenvalue of G_t stays below, and uniformly away from Γ. Hence, $\lim_{t \to \infty} P_1(G_t) = P_1(G_\infty)$. Using Lemma 7.1(3), we have $v_\infty = \lim_{t \to \infty} v_t$. Thus, the conditions of [**IM1**, Thm. V.4.4] are satisfied with corresponding $D = \sum_{k=0}^{\infty} (1 - \epsilon)^k = \epsilon^{-1}$. By the proof of [**IM1**, Thm. V.4.4], we have $P_1(G_\infty) = \lim_{t \to \infty} H(t)$. Lemma 7.1(3) shows that $P_1(G_\infty)$ is fully positive. Thus, we get $v_\infty = \lim_{t \to \infty} w_t$ by applying the argument with the integral once more to $H(t)$ instead of G_t.   □

7.2. Ergodicity for scaled genetic algorithms. Finally, we can state and prove our main result which applies in particular to genetic algorithms using tournament selection but also all other types mentioned above which fit the definition of generalized fitness selection.

THEOREM 7.4. *Let M_μ, $\mu \in [0, \frac{1}{2}]$, describe multiple-bit mutation. Let $\mu(t) \in (0, \frac{1}{2}]$, $t \in \mathbf{N}$, be such that $\lim_{t \to \infty} \mu(t) = \mu_\infty > 0$ exists. Let C_χ, $\chi \in [0, 1]$, be a generalized crossover operation as defined in section 4.1. Let $\chi(t) \in [0, 1]$ be such that $\lim_{t \to \infty} \chi(t) = \chi_\infty$ exists. Let F_t, $t \in \mathbf{N}$, be a generalized fitness selection scaling as defined in section 5.1 with associated $\theta \in [0, 1)$ and $\tau \in \mathbf{N}$. Suppose that $\lim_{t \to \infty} F_t = F_\infty$ exists. Let $G_t = F_t C_{\chi(t)} M_{\mu(t)}$.*

1. *If $v_0 \in S$ is the initial probability distribution over populations for the scaled genetic algorithm, then we have*

$$v_\infty = \lim_{t \to \infty} \prod_{k=t}^{1} G_k v_0 = \lim_{t \to \infty} (F_\infty \mathbf{C}_{\chi_\infty} M_{\mu_\infty})^t v_0$$

   exists and is independent of $v_0 \in S$, i.e., the Markov chain associated with the scaled genetic algorithm is strongly ergodic.

2. *In case of a strong fitness scaling as defined in [**SNF1**, Sec. 2.3] such as power law scaling with $\lim_{t \to \infty} g(t) = \infty$ as discussed in section 5.2, the matrix F_∞ and, thus, v_∞ are independent of the fitness scaling (method).*

3. *If the fitness selection scaling F_t, $t \in \mathbf{N}$, stands for scaled proportional fitness selection, tournament fitness selection in the sense of [**Mit1**, p. 170], [**Gol2**],*

or [**Mic1**, p. 59] *as discussed in section 5.3, or rank selection combined with (scaled) proportional fitness selection as discussed in section 5.4, then the coefficients $<v_\infty, p>$ of the limit probability distribution are strictly positive for every population $p \in \wp$ of uniform fitness. Hence, for a non-constant fitness function, the genetic algorithm G_t, $t \in \mathbf{N}$, does not converge to a population consisting solely of creatures with maximal fitness value.*

4. *If $\wp = \wp_1$ for $\wp_1$ as in the definition of generalized fitness selection scaling in section 5.1, then we have:*

$$\|(\mathbf{1} - P_U) v_\infty\|_1 \leq \theta \frac{(1 - (\mu_\infty^s + (1 - \mu_\infty)^s)^\ell)}{1 - \theta \beta},$$

with $\beta = (\mu_\infty^s + (1 - \mu_\infty)^s)^\ell - ((1 - \mu_\infty)^h \mu_\infty^{s-h} + \mu_\infty^h (1 - \mu_\infty)^{s-h})^\ell$, $h = \lfloor \frac{s}{2} \rfloor$.

5. *Suppose that $\alpha \in (0, 1]$ exists such that $C_{\chi_\infty} = \alpha \mathbf{1} + C^o$, where C^o is a matrix with positive entries. For simple crossover and for unrestricted crossover as in [**SNF1**, Sec. 2.2], this condition is satisfied, if $\chi_\infty \in [0, 1)$. Then we have without the restriction $\wp = \wp_1$ on fitness selection F_t:*

$$\|(\mathbf{1} - P_U) v_\infty\|_1 \leq \frac{(1 - (1 - \mu_\infty)^L)(\tau + 2)}{1 - (1 - \mu_\infty)^{L(\tau + 2)} (\alpha^{\tau + 2} \theta + (1 - \alpha^{\tau + 2}))}.$$

PROOF. Let $H_t = \prod_{k=t}^1 M_{\mu(k+1)} F_k C_{\chi(k)}$. By Lemma 7.2 and Theorem 7.3,

$$\lim_{t \to \infty} H_{t-1} w_0 = \lim_{t \to \infty} (M_{\mu_\infty} F_\infty C_{\chi_\infty})^t w_0 = w_\infty, \quad w_0 \in S$$

exists, and is independent from $w_0 \in S$. Hence, we obtain for every $v_0 \in S$

$$\lim_{t \to \infty} \prod_{k=t}^1 G_k v_0 = \lim_{t \to \infty} F_t C_{\chi(t)} H_{t-1} M_{\mu(1)} v_0 = F_\infty C_{\chi_\infty} w_\infty = v_\infty.$$

v_∞ depends only upon μ_∞, C_{χ_∞}, and F_∞. Using Theorem 7.3 again yields

$$v_\infty = \lim_{t \to \infty} F_\infty C_{\chi_\infty} (M_{\mu_\infty} F_\infty C_{\chi_\infty})^t M_{\mu_\infty} v_0 = \lim_{t \to \infty} (F_\infty C_{\chi_\infty} M_{\mu_\infty})^t v_0$$

This proves (1). (2) follows from the discussion in section 5.2. To prove (3), we observe the following: (CASE 3.1) The case of scaled proportional fitness selection: Let $p \in \wp$ be a population of uniform fitness, i.e., $p = (c_1, c_2, \ldots, c_s) \in \wp$, $c_1, \ldots, c_\sigma \in \mathsf{C}$, and $f_t(c_1) = f_t(c_2) = \ldots = f_t(c_s)$. We have $v_\infty = F_\infty M_{\mu_\infty} C_{\chi_\infty} v_\infty$. Since the p-component of $M_{\mu_\infty} C_{\chi_\infty} v_\infty$ is non-zero, we have $<v_\infty, p> \neq 0$ by Lemma 5.3. (CASE 3.2) The case of tournament fitness selection following [**Mit1**, **Gol2**, **Mic1**]: $F_\infty = F_t$, $t \in \mathbf{N}$, equals the initial fitness selection operator. The probability to reproduce a population of uniform fitness is strictly positive for the methods proposed in [**Mit1**, **Gol2**, **Mic1**] and discussed in section 5.3. Hence, we can argue as in (3.1). (CASE 3.3) In the case of rank selection combined with (scaled) proportional fitness selection as discussed in section 5.4, we can also argue as in (3.1). This completes the proof of (3). To obtain (4), we apply Proposition 3.2(3) to constant matrices $T(t) = F_\infty C_{\chi_\infty}$, and $\mu = \mu_\infty$. To show (5), we observe that

$$M_{\mu_\infty} = (1 - \mu_\infty)^L \mathbf{1} + R,$$

where $\|R\|_1 = 1 - (1 - \mu_\infty)^L$ by [**Sha1**, p. 5, formula (7')]. Using the computation at the end of section 6, we have

$$\|(\mathbf{1} - P_U) v_\infty\|_1 = \|(\mathbf{1} - P_U) F_\infty C_{\chi_\infty} M_{\mu_\infty} v_\infty\|_1 =$$

$$= (1 - \mu_\infty)^L \|(\mathbf{1} - P_U) F_\infty C_{\chi_\infty} v_\infty\|_1 + \|(\mathbf{1} - P_U) F_\infty C_{\chi_\infty} R v_\infty\|_1 \leq$$

$$\leq\ (1-\mu_\infty)^L\,\|(1-P_U)\,(F_\infty\,C_{\chi_\infty})^2\,M_{\mu_\infty}\,v_\infty\|_1\ +\ (1-(1-\mu_\infty)^L)\ \leq$$

$$\ldots\leq\ (1-\mu_\infty)^{L(\tau+2)}\,\|(1-P_U)\,(F_\infty\,C_{\chi_\infty})^{\tau+2}\,v_\infty\|_1\ +\ (1-(1-\mu_\infty)^L)\,(\tau+2)\ \leq$$

$$\leq\ (1-\mu_\infty)^{L(\tau+2)}\,(\alpha^{\tau+2}\theta+(1-\alpha^{\tau+2}))\,\|(1-P_U)\,v_\infty\|_1\ +\ (1-(1-\mu_\infty)^L)\,(\tau+2).$$

Now, (5) is obtained with the argument in the proof of Proposition 3.2(3). $\square$

[**SNF1**, Thm. 18(3)] lists a stronger version of Theorem 7.4(4) for injective fitness functions and strong fitness scalings showing that $\|(1-P_U)\,v_\infty\|_1=0$.

The behavior of proportional selection fitness-scaled genetic algorithms in regard to convergence was posted as an open question in [**Rdo1**]. More generally, one can broaden the question to include genetic algorithms with selection mechanisms based upon rank such as tournament selection, or even other "generalized" fitness selection mechanisms satisfying the requirements in the definition given in section 5.1. This question is answered by Theorem 7.4 in the negative for convergence to an optimal individual or even a population containing one, and in the positive in the sense of convergence to a unique probability distribution over all populations. The main condition is $\mu_\infty>0$. The study in [**GS1**] suggests that in general v_∞ may depend very much upon the fitness function f. However, there is no such dependence on any particular scaling method, if a strong fitness scaling is used. Theorem 7.4 leaves open the case $\mu_\infty=0$ in which the mutation rate converges to zero (an analog of simulated annealing). This question has been addressed and answered for the simple genetic algorithm in the negative by Davis and Principe [**DP1**]. The case of mutation converging to zero in a variation schedule is claimed to be solved in [**Suz1**]. However, in our opinion, there are serious gaps in the presentation of some proofs in [**Suz1**]. Our reservations regarding the validity of the main result of [**Suz1**, p. 66, Thm. 3] are outlined in [**SNF1**, p. 130, footnote 10].

A practical direction for research is to characterize those fitness functions for which genetic algorithms succeed as optimizers in the short-term although our results describe long-term behavior. In fact, the time it is likely to take before an optimal creature appears does strongly depend on the fitness function: In the case with all creatures having small fitness except one with large fitness, iterating the genetic algorithm is no better than blind random search. In fact, in this case, both blind random selection and genetic algorithms are worse than enumerative search, since both will needlessly duplicate choices. Since at each step there is a very small but non-zero chance to choose a maximal fitness creature by mutation, those runs of genetic algorithms which do never choose such a creature have probability zero. Hence, a genetic algorithm will find a creature of maximal fitness with probability 1 in a particular run. This shows that recording a best-so-far creature will optimize the fitness function, but it also shows that in the worst-case scenario this is no better than blind random or exhaustive search. For genetic algorithms to be useful, it is clear that one must have a sufficiently "well-behaved" fitness function.

References

[AL1] Aarts, E.H.L., Laarhoven, P.J.M., Simulated annealing: an introduction. *Statistica Neerlandica* **43** (1989), 31–52.

[AH1] Axelrod, R., Hamilton, W.D. (1981), The evolution of cooperation, *Science* **211**, 1390–1396.

[Bak1] Baker, J.E., Reducing bias and inefficiency in the selection algorithm. In J.J. Grefenstette, ed., *Genetic Algorithms and Their Applications: Proceedings of the Second International Conference on Genetic Algorithms.* Erlbaum (1987).

[CK1] Crow, J.F, Kimura, M., *An Introduction to Populations Genetics Theory*, Harper & Row, New York (1970).

[DP1] Davis, T.E., Principe J.C., A simulated annealing-like convergence theory for the simple genetic algorithm. *Proceedings of the Fourth International Conference on Genetic Algorithms* (1991), 174–181.

[Fog1] Fogel, D.B., Asymptotic Convergence Properties of Genetic Algorithms and Evolutionary Programming: Analysis and Experiments. *Cybernetics and Systems* **25(3)** (1994), 389–407.

[Gid1] Gidas, B., Nonstationary Markov chains and convergence of the annealing algorithm. *J. Statistical Physics* **39** (1985), 73–131.

[Gol1] Goldberg, D.E., *Genetic Algorithms in Search, Optimization, and Machine Learning.* Addison-Wesley (1989).

[Gol2] Goldberg, D.E., *Genetic Algorithms Tutorial.* Genetic Programming Conference, Stanford University, (July 13, 1997).

[GS1] Goldberg, D.E., Segrest, P., Finite Markov chain analysis of genetic algorithms. *Genetic Algorithms and their Applications: Proceedings of the Second International Conference on Genetic Algorithms* (1987), 1–8.

[Hol1] Holland, J.H., *Adaptation in Natural and Artificial Systems.* University of Michigan Press (1975); Extended new edition, MIT Press (1992).

[IM1] Isaacson, D.L., Madsen R.W. *Markov Chains: Theory and Applications*, Prentice-Hall (1961).

[Mic1] Michalewicz, Z., *Genetic Algorithms + Data Structures = Evolution Programs.* (2nd, extended edition) Springer-Verlag (1994).

[Mit1] Mitchell, M., *An Introduction to Genetic Algorithms.* MIT Press (1996).

[MS1] Maynard Smith, J., *Evolutionary Genetics.* Oxford University Press (1989).

[PYW1] Peck, J.R., Yearsley, J.M., Waxman, D., Why Do Asexual and Self-Fertilising Populations Tend to Occur in Marginal Environments? *Mathematical & Computational Biology*, American Mathematical Society, Providence, Rhode Island, (this volume).

[Rdo1] Rudolph, G., Convergence analysis of canonical genetic algorithms. *IEEE Trans. on Neural Networks* **5** (1994), 96–101.

[Rou1] Roughgarden, J., *Theory of Population Genetics and Evolutionary Ecology.* MacMillan Publishing Co., (1976); Reprinted by Prentice-Hall (1996).

[Sen1] Seneta, E., *Non-negative Matrices and Markov Chains*, Springer Series in Statistics. Springer-Verlag (1981).

[Sha1] Schaefer, H.H., *Banach Lattices and Positive Operators*, Die Grundlehren der mathematischen Wissenschaften in Einzeldarstellungen. Springer-Verlag (1974).

[SNF1] Schmitt, L.M, Nehaniv C.L., Fujii R.H., Linear Analysis of Genetic Algorithms. *Theoretical Computer Science*, **200** (1998) 101–134.

[Sig1] Sigmund, K., The Social Life of Automata. In: *Mathematical & Computational Biology*, American Mathematical Society, Providence, Rhode Island, (this volume).

[Suz1] Suzuki, J., A Further Result on the Markov Chain Model of Genetic Algorithms and Its Application to a Simulated Annealing-like Strategy. In: R.K. Belew and M.D. Vose, eds., *Foundations of Genetic Algorithms 4*, Morgan Kaufmann (1997), 53–72.

[Vos1] Vose, M.D., Modeling simple genetic algorithms. *Foundations of Genetic Algorithms* (G. Rawlins, ed.), Morgan Kaufmann (1991), 94–101.

SCHOOL OF COMPUTER SCIENCE AND ENGINEERING, UNIVERSITY OF AIZU, AIZU-WAKAMATSU CITY, FUKUSHIMA PREFECTURE 965-8580, JAPAN
    E-mail address: lothar@u-aizu.ac.jp

CYBERNETICS AND SOFTWARE SYSTEMS GROUP, UNIVERSITY OF AIZU, AIZU-WAKAMATSU CITY, FUKUSHIMA PREFECTURE 965-8580, JAPAN
    Current address: Interactive Systems Engineering, Department of Computer Science, University of Hertfordshire, Hatfield AL10 9AB, United Kingdom
    E-mail address: nehaniv@u-aizu.ac.jp, c.l.nehaniv@herts.ac.uk

Lectures on Mathematics in the Life Sciences
Volume **26**, 1999

Evolution of Diseases

Shinichi Okuyama, M.D., Ph.D.

ABSTRACT. The human genes stock the evolutionary history of life compiled
from prokaryotic to eukaryotic along with hominization. Most of the outdated
genes are preserved but are repressed: Human diseases may arise from a host
of definite evolutionary causes.

(1) Hodgkin's giant cells whose pre-eukaryotic genes are derepressed so as to
overwhelm eukaryotic mitosis (Okuyama 1991); (2) Chediak-Higashi disease
in man has long shared a common lysosomal disorder with the cow and mink
for some sixty million years (Leader 1969); (3) Carcinogenesis in general is
also shared by most of the mammals' protooncogenes (Okuyama and Mishina
1990).

The author now proposes to study evolution of diseases in a more systematic way: We shall overview the matter under several evolutionary principia.
This may constitute a novel view of diseases that would greatly facilitate un-
folding and analyzing the complexity of diseases, and helping to simulate and
reconstruct specific diseases. Then, it will also help us better to understand
diseases and to develop novel diagnostic and therapeutic measures (Okuyama
1997; Okuyama and Sato 1996).

1. Introduction

A host of human diseases can be categorized based on novel evolutionary principia as shown on Table 1. Its needs and rationale will be discussed below along
with an ardent anticipation of its future use in the mathematical and computational
biology and medicine.

With the surge of ever-advancing computer technology, we are the more tempted
to analyze biological phenomena by computer simulation. There are a huge number
of simulations of mammalian cell growth or reproduction. Most of them necessarily
cannot be adequate enough to mimic cell growth: When "cells" come into contact,
they are likely to stop "cell growth" on the instant of contact. In reality, there has
to be a certain latency for those cells to hold multiplication, and the cell contact
inhibition may not necessarily be a physical contact. Therefore, active participation
of clinical physicians and practical biologists is strongly indicated for research into
the biological and computational modelling of human diseases.

The function of a cell, an organ, a system and the whole human body can be
modelled as a black box or a set of black boxes. This type of modelling may have
two merits of investigation: one to get into the heart of the matter by dividing one
black box further into a series of smaller boxes on the one hand, and the other to

1991 *Mathematics Subject Classification.* 92C50, 92D15.

overview the whole picture so as to intergrate definite schemes as soon as certain black boxes are reliably substantiated. Then, our science of mathematical and computational biology may possibly arrive at clues for novel modelling.

The present author tries to illustrate the structure of life so as to facilitate future usefulness in both physiological and pathological modellings. An evolutionary approach has greatly facilitated our better understanding of various complex diseases such as cancer (Okuyama and Mishina 1990) as well as the "normal" complex structures and functions of organs as was the case with cognition (Okuyama 1997): a series and/or a hierarchy of cognitive black boxes. Such an evolutionary approach will help us decipher the layered architecture and function of the modern cells, organs and systems, and the whole host animal. This approach may be one of the ways to analyze and characterize the black boxes of interest.

It should be noted here that a biological system is not only a physicochemical construction but it also possesses innate motivation to live that drives its animal host along the stern food chain and that impels a plant to secure light and nutrient chemicals to sustain itself. Thus, most of the biological functions are understood as being purposefully driven.

A disease is (1) a definite morbid process having a characteristic train of symptoms; (2) affecting the whole body or any of its parts; (3) its etiology may be genetic or non-genetic: non-genetic etiology includes infections, extrinsic noxa, intrinsic derangements or simple oscillations of the *internal milieu*. Thus, a disease can be defined as a black box or boxes causing morbidity of the patient with not necessarily any perfect regards to the proven etiological cause-effect rules.

From a radiological point of view, a cancer cell is large, decreased of superoxide dismutase, and defective in DNA repair (Fig. 1) (Okuyama and Mishina 1984). Then, the evolutionary concepts of human diseases struck me when I thought how

I. Diseases relative to endosymbiotic structural evolution:
(A) Nuclear: Chromosome breakage syndrome; Fanconi's anemia;
   Xeroderma pigmentosum; Cancer;
(B) Divisional apparatus: Down syndrome; Reed-Sternberg's
   giant cells in Hodgkin's disease;
(C) Mitochondrial: Mitochondrial encephalomyopahty;
(D) Lysosomal: Chediak-Higashi syndrome;
(E) Ciliary: Primary mucocilliary transport failure;
II. Hominization disorders or Diseases relative to hominization:
(a) Disorders relative to bipedality; Perthes' disease;
   orthostatic dysregulation syndrome; spinal disc hernia;
   cerebral vascular diseases; cardiac vascular disorders;
(b) Immune hominization resulting in the generation of atopy,
   bronchial asthma, pollinosis;
(c) Disorders relative to the one brain-two hemispheres
   doctrine: motor and sensory aphasia;
(d) Hominization disorders of the specific sensory organs:
   Olfactory atrophy leading to less dependency of *Homo sapiens sapiens*.

TABLE 1. **An Evolutionary Outlook on Human Diseases**

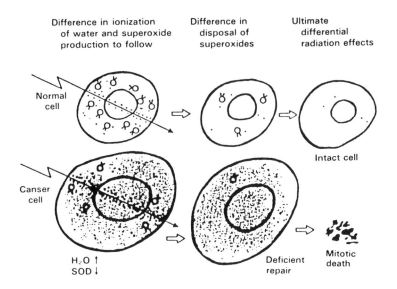

Difference in ionization
of water and superoxide
production to follow

Difference in
disposal of
superoxides

Ultimate
differential
radiation effects

Normal
cell

Intact cell

Canser
cell

$H_2O \uparrow$
SOD $\downarrow$

Deficient
repair

Mitotic
death

FIGURE 1. **Origin of the selectivity of radiotheraphy of cancer.** We know that when we irradiate a cancer mass in the deep chest, we can dissolve that mass. Even then, however, the skin, muscles and lung parenchyma overlying will not dissolve: a selective effectiveness of radiotherapy. Cancer cells are different from normal cells not only in their cytosomal largeness but also in terms of increased water density, decreased content of superoxide dismutase (a superoxide scavenger), and a reduced capacity of excision repair of DNA damage. Because of this triad, the author was struck by an evolutionary notion of cancer (Okuyama 1996; Okuyama and Mishina 1990). Reprinted from *Evolution of Cancer* by S. Okuyama and H. Mishina, University of Tokyo Press, 1990, by permission of the publisher.

to improve radiotherapy: A cancer cell in reality represents a state of devolution with an increased water content, decreased DNA repair and decreased oxygenation (Fig. 2) (Okuyama and Mishina, 1984, 1990). The legend of this figure well explains mysteriously complex attributes of cancer. When he wrote about cancer devolution, Setala was right in ascertaining that cancer may arise at the *loci minores resistentiae* (weak points of resistance) of symbiotic junctions (1984). However, he was too old to expand his concept to those non-cancer diseases.

We have first of all tried hard to incorporate evolutionary concepts in the elucidation of several non-cancer diseases as well. We have secondly adopted Hunter's arterial circle (Brookes 1971) as one of the critical principles of organ perfusion in order to study the coronary obstruction and cerebral vascular disorders.

Thus, we may assume that the present information might possibly encourage future mathematical and computational simulation of the biological and pathological phenomena.

With ever-increasing and ever-widening medical knowledge, especially that of genetic and molecular biology along with that of biochemistry, a student of medicine or biology, and possibly of mathematical and computer science will certainly expect

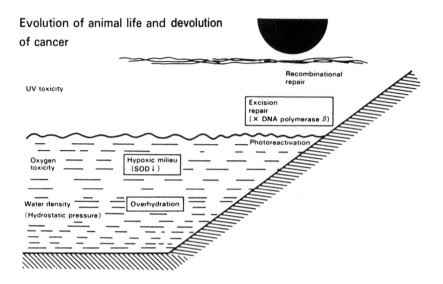

FIGURE 2. **Evolutionary concept of cancer.** The triad cancer attributes led us to the idea that "cancer" mimics life in the deep primitive sea with reduced penetration of UV light, reduced oxygen diffusion and increased hydraulic pressure. To increase intracellular osmotic pressure, almost everything is dissolved in the cytoplasm rather than being packed in the organellae (Okuyama 1996; Okuyama and Mishina 1990). Reprinted from *Evolution of Cancer* by S. Okuyama and H. Mishina, University of Tokyo Press, 1990, by permission of the publisher.

to have it made more concise and more comprehensible. Then, hopefully, the present exposition shall be of some help.

2. Principia of Evolution:

Evolution can be discussed in two different categories: (1) phenotypic, and (2) genotypic. It is represented in a process of development in which an organ or organism becomes more and more complex by the differentiation of its parts. The essential change must take place in the genes. However, it should also be noted that such change has to take over the gene pool of a population over generations by such processes as mutation, natural selection, and genetic drift. From our purpose of deciphering evolutionary disease puzzles, the following principia will be applied.

A. Principium of Metamerism and Metamorphosis. The bodily shape and function of the vertebrate animals has long been controlled by the metamerism and metamorphosis. *Metamerism* denotes the arrangement into somites or metameres by the serial repetition of a structural pattern such as the vertebral bodies of the vertebrate animals. *Metamorphosis* implies a complete change in appearance or character or else of the whole or parts of the body. Metamerism is originally 'intended' to procure stable and durable body plans. Metamorphosis, on the other hand, is constant 'endeavor' to better adapt the animals in question to the novel changes in the food chain and ecological environment. The former is integrative while the latter is differential adaptation. Homeobox genes or a group of genes control the relevant competent cells to migrate, divide and/or mature and to form organs or tissues: While embryonic branchial arches r3 through r6 design the shape and structure of the oropharynx, r1 and r2 form the maxilla and mandible. The formation of the vertebral column is the function of metamerism. Recent studies have confirmed that the formation of the oropharynx is again metamerism with slight metamorphosis. The jaws, on the contrary, are rhombomeres or the metameres derived from r1 and r2 in the rhomboencephalon. That derivation of jaws from the metameres was a long evolutionary history: The fish, amphibian, and reptiles do not possess true dentitions: Their teeth are cartilaginous.

A′. Biochemical Metamerism and Metamorphosis. Having in mind the intention of expediting decipherment of evolutionary mystery and arriving at evolutionary *loci minores resistentiae*, the definition of metamerism and metamorphosis can be expanded into the domain of biochemical evolution.

A photosynthetic plant cell is a metamere of photosynthesis, and therefore, a plant represents a biochemical metamerism as it is composed of numerous, similar photosynthetic cells.

An animal cell is likewise a metamere of glycolysis for energy for activity of whatever sort. Then, an animal represents a biochemical metamerism of glycolysis.

Biochemical metamorphosis is well exemplified by a different evolutionary treatment of purine wastes: With the fish, the nitrogenous wastes are oxidized to ammonium which is dissipated into the surrounding water without leaving any ammoniac toxicity behind. The biochemical metamorphosis of land amphibians is to produce urea. That of *Homo sapiens sapiens* as well as the higher primates is to produce urate which is least water-soluble, but which is highly likely to produce gout or urate crystallization inside body tissues only to result in painful tophous formation. The birds secrete urate into the cloaca in which urate is mixed with feces and evacuated without any urate depositions.

B. Principium of Heredity: Life is defined as an organism that descends from its ancestor, that is capable of sustaining itself by procuring energy and nutrients from its surroundings, and that is capable of reproduction itself. The information for such setups has been inherited from one generation to another over the past period of 4 billion years. Every life is fitted to the food chain: it exploits the weaker but is likely to become the prey of its natural enemies. It has tried to exploit the most of its prey on one hand, and it has tried to be exempted from, or to acquiesce of its predators on the other. Then, the life has been driven to pursue prey and avoid its predators as much as possible.

In order to keep up with the food chain, every life had to maintain its bodily structure and function: a strict adherence to heredity. In order to conserve its genetic information, it had to repair those errors incurred by extrinsic and intrinsic noxa such as UV light and active oxygen molecules arising within. Unrepairable errors could have contributed to the emergence of mutations. As novel mutations were found useful and prevailed, much of obsolete genetic information could have been repressed. Thus, as soon as the novelty was transmitted from one individual life to another and fixed in that species, the change was an evolution. Evolutionary *loci minores resistentiae* may arise when the hereditary genetic information fails to be conserved in one way or another.

Genes or heredity units of a cell are assemblies of nucleotides, special chemicals, that function as information for cell physiology and multiplication. With the eukaryotes, these units are packed in the nucleus while they are not with the prokaryotes. During the process of cell division, these genes are seen to be grouped and housed in many chromosomes. In these chromosomes, they are arranged longitudinally. Each chromosome is banded. The hereditary units will function individually. Nonetheless, it is noteworthy that the units represent metameres and the entire arrangement is metamerism. If we imagine a **zootype** or prototype animal or a **phytotype** plant, any evolutionary transformations will represent metamorphosis.

C. Principium of Horowitz's law of biochemical evolution: According to the law of biochemical evolution of Horowitz (1945), the evolution of the basic syntheses proceeds in a stepwise manner, involving one mutation at a time, but the order of the attainment of individual steps was in the reverse direction from that in which synthesis proceeds in the chain. The ultimate synthesis was the first to be acquired in the course of evolution, the penultimate step next, and so on.

This principle firstly helped us deciphering the mystery of the biochemical chain of glycolysis (Nakamura 1982; Okuyama and Mishina 1990). Figure 3 illustrates this principle: the biochemical processes of tri-carbons that are capable of generating ATP and NADH without any pre-investment of ATP. Compare them with the breakdown of hexoses such as glucose and fructose which requires ATP pre-investment. Thus, the tri-carbons could have been the ultimate synthetic step, and

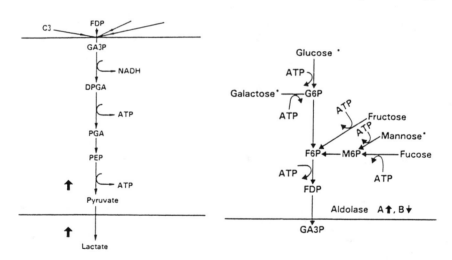

FIGURE 3. **Principium of Horowitz's law of biochemical evolution: The primitive fermentations.** A. The tricarbons (C3 and FDP) are processed to yield pyruvate without any pre-investment of ATP molecules during which catabolism the energy coins of ATP and NADH are produced. B. To the contrary, ATP pre-investment is mandatory in order to have hexoses (glucose, galactose, fructose, mannose and fucose). Then, the synthesis of pyruvate from the tricarbons is supposed to be the first step of acquisition (Horowitz 1945, Nakamura 1982, Okuyama and Mishina 1990). Reprinted from *Evolution of Cancer* by S. Okuyama and H. Mishina, University of Tokyo Press, 1990, by permission of the publisher.

the chain of reactions from hexoses to the tri-carbons, the penultimate synthetic processes.

In the primitive seas, tri-carbon organic substances could have been rich enough to sustain the primitive cells until they were exhausted and photosynthetic green algae radiated.

D. Principium of Endosymbiotic Loci Minores Resistentiae: As mostly accepted, the present eukaryotes or cells other than bacteria could have been generated by endosymbiosis of prokaryotes (Margulis 1981). Resultantly, there can be disorders at sites of symbiotic junctions as wells as in those of incorporated nuclear and cytoplasmic structures themselves.

E. Principium of Hominization: Hominization, that is, the evolution of human characteristics different from other primates, certainly represents the gorgeous achievement of many human traits such the large, competent but asymmetric brain. The human brain is a hybrid of *reflex* and *thinking* brains. The reflex brain is a hierarchical system consisting of a serial neuronal ascending connection reaching the sensory centers on one hand, and a similar but descending connection from the motor centers on the other. Studies of aerial mapping of neurons in the motor and sensory strips prove interconnections between them. Nonetheless, as soon as extensive cognitive search and analysis come into operation with the association cortices in such tasks as listening to speech, there emerges hominized thinking. The human brain is asymmetrical from right to left; the speech centers localize in the left hemisphere, spatial cognition and musical enjoyment in the right. Its largeness comparative to the body weight is estimated by the psychic factor of Snell [$h = p \times k^{\frac{2}{3}}$ where h stands for the brain weight, p for psychic factor, and k for the body weight] (1982). It is as large as 0.97 as compared with 0.12 for a clever lion, 0.18 for an African elephant, and 0.34 for a chimpanzee (Fujita 1988). A brain as heavy as that of a man cannot be sustained in a horizontal plane with a quadrupedal animal. It thus necessitates bipedality. Nourishment of this large brain is also a problem. The cerebral circulation is so modified as to minimize the capacitor function of the arterial circle of Willis in order to acutely perfuse the active portions of the brain. This is blood supply on demand. This design helps economize and presumably reduce active oxygen toxicity in the cerebrum.

With such a brain, then, human cognitive capability greatly became keener and more accurate than in other primates: acuity, binocular vision, and color vision.

In order to make full use of the cerebral neurons, the human cerebral hemispheres are allotted jobs different between the right and left: This same asymmetry of the brain, however, may entice untoward hemiparesis in man when ischemic events take over the organ, leading to unilateral sensory and motor impairment, aphasia, and dementia.

The immune system has also been hominized by the evolution of IgE (Ishizaka and Ishizaka 1966). This is concerned with the immediate type hypersensitivity. It functions as a precipitor of the relevant antigens in the skin by binding to receptors on mast cells and letting them degranulate and secrete biologically active substances that are responsible for development of atopic reactions.

F. Principium of Hunter's arterial circle as the prototype vascular capacitor: We should well be surprised by Nature's art of a beautiful body of an exact symmetry. What is its secret? Our conclusion is Hunter's arterial circle

(HAC) (Fig. 4) (Brookes 1971). It is a vascular circle that receives two or three afferent arteries from the systemic circulation and sends off efferent arteries to nourish the epiphysis for growth and metaphysis of a long bone.

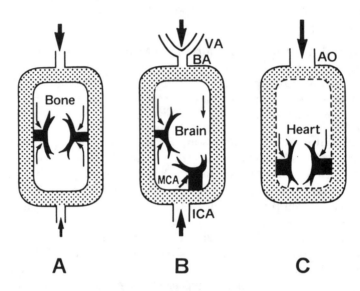

FIGURE 4. **Principium of Hunter's arterial circle as the prototype vascular capacitor.** A. Hunter's arterial circle (HAC) in the bone is supposed to be one of the prototype vascular capacitors. This model facilitates understanding of the burden on bones for bipedal human evolution in regard to Perthes' disease in boys and hyperartrophic oseoarthropathy in infants treated with vascular dilators such as prostaglandin E1 (Okuyama 1992; Tomiie *et al.* 1982). B and C. Expanded capacitor models for the brain and heart: Willis' arterial circle for the brain is an attempt at omitting the capacitor in order to ensure a prompt, on-demand blood supply during diastoles (Okuyama 1992). The alterations represent specific adaptations to hominization of these specific organs respectively. Reprinted from "Impacts of the evolutionary concepts of cancer on the study of human diseases" by S. Okuyama in *Tohoku Journal of Experimental Medicine* (168), pp. 446, 1992, by permission.

HAC functions so as (i) to compensate if one of the afferent arteries fail or (ii) to moderate rises and falls of the systemic blood pressure or (iii) to generate gradient in blood pressure or (iv) to diminish arterial impact by means of energy dissipation during passage through the elastic pipe a certain geometric distance.

HAC's capacitor function may arise from (1) an added space for the blood to spread, (2) Bernouille's depressor effect, (3) vascular mural elasticity to bounce to dissipate energy, and (4) the pressure gradient between pre-HAC and post-HAC circulation.

The principium of HAC is one of those of hominization. Here, it should also be noted that a further localization-wise subcategorization of the principium into pre-HAC, HAC itself and post-HAC may probably be of great help in better understanding disease mechanisms.

G. Principium of mosaicism. In our body, mosaicism between prokaryotism and eukaryotism may exist especially in disease states such as in the case of Hodgkin's disease to be discussed later. Within a cancer tissue, there are hyperemic or hyperoxic regions along with hypoxic areas even though the latter do not represent necrotic or necrobiotic tissues. There can be hypermetabolic areas coexistent with malnutritious. At this moment, we are not to have any 100% primitive tissues or biochemical functions. Hodgkin's disease is another example. It is a disease of pre-eukaryotism because of amitosis in a selected population of cells of the body. However, in a group of patients of the same disease of nodular sclerosis type, their pathological cells do not exhibit any signs of that pre-eukaryotic cell division. Mosaicism may be existent but temporarily in inflammation, on-going fibrosis, and so on, too.

3. Pre-Hominization Diseases

From an evolutionary point of view, human diseases as a whole can be discussed in two distinct categories: those of hominization and those of pre-hominization (Table 1). Selected diseases will be discussed here.

Hodgkin's disease. In the early part of the 19th century, Thomas Hodgkin first described a disease of systemic and fatal lymph node swelling (Hodgkin 1826). Although he could not carry out any microscopic studies then, his insightful observations are still valid. It is characterized by giant cells of bizarre configuration called Hodgkin and Reed-Sternberg cells. Their origin and morphogenesis are still unsettled.

When I examined histopathological materials of a patient of Hodgkin's disease, I came across a bizarre four-nucleated giant cell. It was not carrying out any

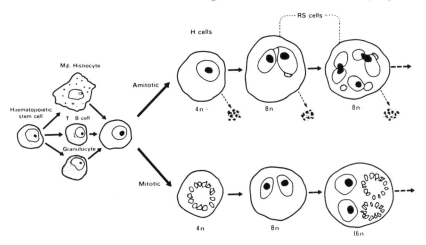

FIGURE 5. **Generation of Hodgkin and Reed-Sternberg giant cells (H and RS cells).** They are produced through amitosis as well as mitosis. Amitotic H and RS cells are more likely to disintegrate in due course of time than those of mitosis (Okuyama 1996). The change may represent the principium of mosaicism as well as cancer devolution. Reprinted from "Evolution of Hodgkin's disease" by S. Okuyama in *Med. Hypotheses* (47), pp. 199–213, 1996, by permission of the publisher Churchill Livingstone.

mitosis but all the four nuclei are stretched out so as to be eventually torn apart. The picture reminded me of a Euglena cell, a primitive eukaryote, undergoing cell division (Okuyama 1991). It was amitosis rather than mitosis! Now we know that even the selected human cells may undergo amitosis: budding (Stobbe 1981) and amitosis (Okuyama 1991). In a detailed histopathological study, division of giant Hodgkin cells were traceable to transform into Reed-Sternberg cells through an equal, amitotic nuclear division (Fig. 5) (Okuyama 1996). How can a genuine human cell undergo amitosis? Our conjecture at the moment is as follows: A Hodgkin cell stocks pre-human genetic information that is repressed. Portions of the stock can be derepressed so as to express themselves in amitosis.

The origin of cells of Hodgkin's disease has long remained unsettled — monocytes or myeloid cells or lymphocytes or other hematopoietic stem cells? When we think of the Hodgkin's patients arising in the host of people exposed to certain chemicals and those having a history of infectious mononucleosis due to an EV virus and showing an invariable histopathological nodular sclerosis, we had better advise ourselves to acknowledge that any kind of hematopoietic stem cells will develop Hodgkin's disease should the chemical and/or viruses derepress the amitotic genes and interfere with mitosis in the selected, susceptible cells. Peculiarly enough, Hodgkin's giant cells seldom come into close contact with each other and they do not pile up as in the case of cancer cells. Rather they are separated from one another and perish in the shape of "lacunar formation" (Okuyama 1996). Apoptosis (Okuyama and Mishina 1990) or not, the giant cells do not proliferate at a faster speed but perish in the lacunae.

Other cytogenetic evidence supports that the nodular sclerosis type of the disease is derived from mitotable cells of viral immortalization and transformation (Okuyama 1996). Then, Hodgkin's disease embodies the principium of mosaicism in two different levels: (i) cellular and (ii) host as a whole.

The disease does not appear like a genuine neoplasm. It is also characterized by production of cytokines such as IL-6. The latter may be responsible for its inflammatory pleomorphism as eosinophilia throughout the pathological tissue. Nonetheless, it remains one of the pre-hominization disorders.

Cancer. Cancer represents devolution of contemporary normal cells to a "primitive animal life" in which cancer cells multiply slowly under hypoxic and hyponutrient milieu (Sasaki *et al.* 1981, Okuyama and Mishina 1990). As shown on Table 2, the ascitic cancer cells can grow under practical anoxia. They grew despite the nil glucose ascites. They may live on amino acids such as glutamate instead, however.

	pH	pCO2 (mmHg)	pO2 (mmHg)	Glucose (mg/dl)
Blood	7.3	53.0	43.4	120
Ascites	6.8	111.4	0.35	2.5

TABLE 2. An experimental ascitic cancer of the rat AH109A live under "anoxia" without appreciable glucose. The ascites were obtained for analysis during log phase *in vivo* proliferation (Okuyama 1992).

Cancer is a metamorphosis due to changes in the genetic metameres in somatic cells through activation of protooncogenes or generation of novel carcinogenic genes.

Tumor Incidence Rates(%) in Japan : An Evolutionary View

			MALE Calendar Age								FEMALE Calendar Age							
			0~4	10~14	20~24	30~34	40~44	50~54	60~64	80~84	0~4	10~14	20~24	30~34	40~44	50~54	60~64	80~84
Evolutionarily Secured		Testis Ovary																
Evolutionary Age — Vertebral	Mammalian symbols	Breast									0.9	0.3	6.4	22.0	25.0	17.3	10.6	5.8
		Uterus									0	0.3	11.5	17.4	33.1	23.8	15.8	6.4
		Prostate	0	0	0.3	0.3	0.1	0.3	1.8	5.3								
		Bladder	0	0	0.3	1.4	1.7	1.7	2.6	4.7	0	0	0.3	0.1	0.1	0.8	1.0	1.9
	Homeothermal	Thyroid	0	0.4	1.2	0.9	0.1	0.2	0.3	0.2	0	2.8	7.3	3.3	1.8	1.8	1.3	0.8
		Larynx	0	0	0	0.2	0.3	1.0	2.1	1.5	0	0	0.1	0	0.1	0.1	0.2	0.2
		Lung	0.6	0.3	2.0	3.2	5.5	9.2	12.5	12.1	0.2	0.8	0.6	1.1	2.2	4.4	5.8	7.3
	Poikilothermal	Esophagus	0	0	0.4	0.3	1.1	3.6	4.9	4.7	0	0	0	0.1	0.1	0.9	1.4	2.3
		Stomach	0	0.5	14.3	37.3	47.4	47.7	42.8	40.0	0	1.5	20.5	33.6	25.6	24.6	31.7	34.4
		Colorectum	0	0.9	4.6	12.2	10.2	8.0	8.5	8.9	0	0	2.5	6.6	5.9	7.8	8.9	12.8
		Hepatobiliary	2.8	1.3	3.5	3.8	10.1	10.8	10.5	9.5	7.3	6.4	2.1	3.6	2.0	5.9	7.0	12.2
		Pancreas	0	0.5	1.0	0.7	2.1	3.1	3.3	2.2	0	0	0.7	0.3	1.2	1.7	4.0	3.2
		Brain	12.1	20.8	8.0	2.8	1.6	0.6	0.5	0.1	14.2	15.1	3.2	3.1	1.6	0.8	0.6	0.1
		Kidney																
		Bone	3.1	6.8	3.4	0.7	0.5	0.3	0.3	0.3	0.5	12.1	1.9	0.4	0.3	0.3	0.2	0.2
	Pre-vertebral	Lymphomas & Leukemias	39.2	47.1	36.5	18.3	7.8	5.5	3.8	3.4	42.4	37.7	11.8	4.4	2.0	2.2	1.2	0.7

FIGURE 6. **Tumor incidence rates in Japan: An evolutionary view.** When the organs of tumor origin were arranged according to their evolution on the earth, pre-vertebral and vertebral, and latter further subclassified into poikilothermal, homeothermal, and mammalian, there appeared a defined trend of tumor emergence that increased with aging. During the early periods of life, one is liable to develop malignancies of the more primitive life, resulting from a small number of gene alterations for carcinogenesis, say, one or two or three for leukemias. Conversely, several genetic changes have to be produced for the older people to develop cancers. Carcinogenesis itself is a conflict between the induction of carcinogenic DNA changes and subsequent growth of abnormal cells on the one hand and the defence lines of immune surveillance and apoptosis (*horror autotoxicus*) on the other. The data on tumors from the testis and ovaries were too low to be listed in the table. This finding was thought to imply that these organs may possess mechanisms within themselves to eliminate cells of abnormal genetic information (Okuyama and Mishina 1990).

Carcinogenesis progresses from immortalization to transformation. This process is well analyzed in the development of colon cancer in those candidate patients having intestinal polyposis.

Cohort studies on natural carcinogenesis in man revealed that with advancing age, the prevalent malignancies replay the evolutionary history of animal (Fig. 6): (1) Leukemias or unicellular tumors for the youngest; (2) tumors of the bones and kidney or mesenchymal organs; (3) tumors of the central nervous system; (4) tumors of the gastrointestinum and liver or organs from poikilothermic animals for the adult; (5) tumors of the homeothermic organs such as lung and thyroid for those of middle age; and (6) tumors of the breast and uterus, and prostate (male uterus) symbolizing the mammalians for the elderly or post-reproductive populations (Okuyama and Mishina 1986; 1990).

In addition to these "basic" cancers, there are "variable" cancers that will emerge along with the social and environmental trends of viral, chemical, earthly and atmospheric contamination (Sugano 1980; Okuyama and Mishina 1990). They are (a) stomach cancer for the young and middle-aged adults representing chemical carcinogenesis from the food additives; (b) breast and uterine cancers for women of the reproductive period from viral and endocrine stimulation; and (c) prostatic cancer for the men of reproductive period from food or viral or racial reasons (Okuyama and Mishina 1990).

Identical trends were observed in the results from the analysis of cancer incidence in the population of Nagasaki City 30 years following the atomic bombing in 1945 (Fig. 7) and cancer incidence among the patients of Fanconi's anemia (Okuyama and Mishina 1987, 1988). With atomic bombing, radiation effects of some kind could have been causative while with Fanconi's anemia, an increased superoxide radical circulation from deficiency in superoxide dismutase could have been responsible. These are concerned with production of niches on DNA molecules. Some of the unrepaired niches are responsible for carcinogenesis.

Most protooncogenes so far discovered are widely distributed throughout the animal kingdom. They are not only shared by the mammals but also by other vertebrates such as fish and birds. Their origin is the vital genes concerning cell proliferation or growth regulation: DNA, reverse transcriptase type RNA or suppressor genes. Therefore, we may speak of "cancer of the whole animal kingdom in one language." We may further discuss the evolution of cancer as it progressed from (i) DNA defective type (Leukemias) to (ii) DNA repair type (epithelial cancers), (iii) immune exaggeration (lymphomas or tumors of the lymph nodes), and (iv) viral carcinogenesis (DNA, RNA and retroviruses) (Fig. 8) (Okuyama 1996; Okuyama and Mishina 1990).

The lymphatic system has evolved over the past hundred million years: Lymphocytes first and lymph nodes later. Viral carcinogenesis is thought the most novel because those viruses causative to a species may have been derived from that species or its nearest ancestors.

Thus, cancer has been the constant product of animal life, and it has become more complex with time presumably as the animals evolved ways of defence against against carcinogenesis for themselves: (i) DNA repair, (ii) superoxide dismutase, (iii) immune surveillance, and (iv) apoptosis (Okuyama and Mishina 1990).

Then, carcinogenesis is a battle of the host to defend its metameric structure and homeostasis until it yields to cancerous metamorphosis.

Radiation dermatitis and mitochondria. In mitochondria, the terminal portions of exploitation of energy in terms of ATP molecules from fractionated glucose are being carried out.

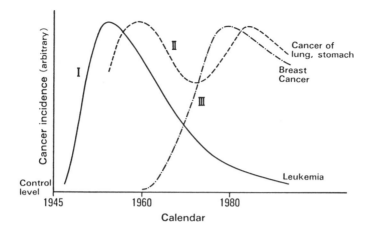

FIGURE 7. **Atomic bomb radiation carcinogenesis in the population of Nagasaki city 30 years after the bombing:** Leukemias (I) to epithelial cancers (II) and to cancers of the mammalian symbol organs (III). The women of reproduction period could have migrated into the city until they submerged into those of post-reproductive phase (Okuyama and Mishina 1988, 1990). The trend was a replay of Nature's carcinogenetic experiment observed in a cross sectional study (Fig. 6). Reprinted from "Cancer incidence in the population of Nagasaki City 30 years after the atomic bombing" by S. Okuyama and H. Mishina in *Tohoku Journal of Experimental Medicine* (155), pp. 39, 1988, by permission.

Radiotherapy damages the mitochondria. Such damaged mitochondria will recover if sufficient amounts of cytochrome c are dispensed. On completion of the radiotherapy to 100 Gy, the skin of the radiation field was red in color and warm and tender to palpation. Nonetheless, as soon as cytochrome c preparation was commenced on that patient, erythema, swelling and tenderness were all gone (Okuyama and Mishina 1982; 1990).

This cytochrome c effect was microscopically confirmed (Okuyama and Mishina 1990). All the characteristics of the untreated control radiation dermatitis capillary dilation with red cell engorgement and intense blue coloration from extensive collagenous precipitation improved on cytochrome c treatment.

Thus, nowadays, we give radiation doses at least 10 to 25 % greater than the conventional doses. The administration of cytochrome c from the beginning of radiotherapy helps substantially reduce the degree of radiation dermatitis and colitis.

Lysosomes. Lysosomes are descendants of a symbiont whose genes are retained in the nuclear genes (Margulis 1981). Chediak-Higashi syndrome is a disease of the lysosomes in the cells that function as a digestive structure for bacteria and other particulate materials and organisms. The anomalous lysosomes cannot fight with bacteria, and children having this disorder may not survive beyond the 6th or 7th year of life. Leukocytes are expected to contain delicate granule-like lysosomes throughout the cytoplasm. With Chediak-Higashi anomaly, however, the lysosomes are extraordinarily large and they cannot function as champions to fight

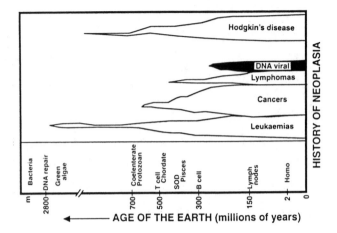

FIGURE 8. **Evolution of cancer.** Neoplasms in multicellular life evolved from simple leukemias into cancers and other malignancies along with the evolution of facets of the lymphatic immune system: natural killer (NK) cells, T cells, B cells, and formation of lymph nodes. The neoplastic phenomenon has pre-hominization history. Reprinted from "Evolution of Hodgkin's disease" by S. Okuyama in *Med. Hypotheses* (47), pp. 199–213, 1996, by permission of the publisher Churchill Livingstone.

the invading bacteria (Wintrobe 1981). According to Leader (1969), this lysosomal anomaly has been long shared by the cow, mink, mouse and man for some sixty million years.

Thus, we have learned that many human diseases may have a close connection with the pre-hominization or evolutionary history of animal life.

4. Hominization Diseases

Bipedality and encephalization, both symbolize hominization. The human cerbral hemispheres represent a "functional metamorphosis" as well as a morphological one. Both features are deeply concerned with the subsequent disease sequelae as will be discussed below.

The brain originally is a metamorphosis from the spinal cord metameres. Human encephalization or the formation of a large human brain is an evolutionary sequela of hominization: In order to better survive the harshness of the food chain, mankind had to develop the cerebral association cortices along with the superior cognitive organs. Hominized vision, for instance, is characterized by (1) binocular vision with a wider visual field and stereopsis; (2) great visual acuity and depth perception; and (3) color vision. The visual information as it leaves the primary visual cortex will be transferred to the association cortices which function as an organ of analysis and judgement. For the sake of prompt and comprehensive function, there has to be a great number of neurons and their processes interconnecting with each other. Thus, an enormously large brain is human.

Bipedality is also human. To be bipedal is to erect oneself. Excessive bodily burdens, acute bending, and muscular contraction will produce lumbar and cervical disc herniation. Likewise, in selected susceptibles, a combined organovascular alteration on erect posture may give rise to syncopes. Firstly, these will be discussed as representative hominization disorders. Secondly, we have to know that to keep oneself in an erect posture, the heart has to pump up the blood some 30 cm higher than in the case of quadrupedalism. This is quite a load, and it affects the elderly appreciably (Shannon *et al.* 1992).

Probable immediate demands of evolved bipedality in man were to weigh the entire body with the two hind limbs and yet to permit a quick and stable onward gait. The pelvic and hip joints were remodelled: (a) The pelvis became shallower but wider; (b) the lower extremities were adducted; (c) the gluteal girdle was well developed so as to firmly suspend the lower limbs; (d) the quadrupedal animals walk or run with a circumflex swinging of the limbs in the Mann-Wernicke type motion. The bipedal swing in man is different. As post-apoplectic patients express it, the Mann-Wernicke motion is more archaic than the bipedal swing. The hemodynamic load upon HAC in the hip joint could have consequently increased more than twofold, probably by a factor of 2.5 as suggested by dynamic analysis of the hip joint (Greenwald 1991).

Lumbar disc hernia. The long, horizontal vertebral column of a wild quadruped animal provides a predator with a *locus minores resistentiae* of the prey animal. A strong blow in the lumbar region would certainly break up the spinal cord, paralyze the prey from spinal dissection and subsequent development of fatal spinal shock.

Disc herniation is common in the cervical and lumbosacral vertebral columns. The herniated nucleus pulposus impinges upon the spinal cord, producing severe pain, numbness and paralysis. In order to cope with the burdens of the heavy head and body in erect posture, the nuclei pulposus are intervertebrally inserted to function as cushions. The nucleus has to be both soft and stiff enough in order to buffer gravitational and untoward stressings.

Thus, both lumbar and cervical disc herniation are hominization disorders. The heavy head and trunk will compress the intervertebral discs, especially when an acute bending or jump and falling squeeze the discs to rupture and extrude the nucleus pulposus. The latter will tease the nerves and nerve roots to cause neck pain and lumbago, num the nerves and not infrequently paralyze the patients. Thus, disc herniation is literally a hominization disorder.

Orthostatic dysregulation syndrome: Nausea and fainting may strike mostly girl pupils when they are forced to keep standing for a certain period of time. One or both of the kidneys may fall spontaneously in such patients. A different study has revealed that vascular, and therefore, hemodynamic change will follow this organoptosis. A renal angiography in an erect posture of an adult patient demonstrates that the ptotic kidney fell deep into the pelvis, dragging the right renal artery downward. Then, the aortic blood will flow into the renal artery without being curved at right angle and the receiving kidney will be stressed in various ways. The hemodynamic changes may be identical with those to accompany the mictrition syncope in a young but tired man who forces urination following excessive beer and all of a sudden faints. In the young children, their growth in body length far exceeds that of weight, and as soon as their body weight catches up with the growth in body length fainting spells disappear.

Skeletal system. The principium of Hunter's arterial circle (HAC) will be challenged in two extreme diseases: (1) Perthes' disease in growing boys, and (2) secondary hypertrophic osteoarthropathy of infants.

Perthes' disease. Perthes' disease is an idiopathic aseptic necrosis of the epiphysis of the femoral head in growing children, mostly boys (Barnett 1957). Vascular anomalies are behind the disease (Sevitt and Thompson 1965).

In bipedal walking, shearing force is generated across the caput and collum femoris as evidenced by the fact that the primary lamelae are laid down parallel to the lines of stress (Harty 1991). This would break up the delicate balance of osteoclasis and osteoplasia in the growing bones or interfere with healing of the traumatic fractures. HAC now consisting of the medial and lateral circumflex femoral may not always be sufficient enough in defraying the increased demands of blood supply even in the absence of any obvious traumas to the surrounding soft tissue. With the athletes, formation of abundant collateral circulation is probable, and this would help in diverting aseptic necrosis of the femoral head. Thus, Nature seems to torture bipedal man by shearing the femoral head as well as by imposing a greater hemodynamic burden upon HAC in the hip joint. Because a greater blood supply is needed, the number of afferent arteries and/or a circle of greater diameter should be evolved in order to increase the magnitude of the circle's capacitance. A second way can be to encourage formation of collateral circulations.

In the growing children, the circles have to cope with the incessantly increasing demands of blood supply along with the bipedal metabolic increase of evolutionary origin. Thus, Perthes' disease is a primary HAC disorder in which blood flow to the growing articular bones is deficient. It is one of the hominization disorders as it arises with the evolution of bipedalism, and constitutes one of the evolutionary *loci minores resistentiae.*

Secondary hypertrophic osteoarthropathy of infants. In those newborns suffering from congenial cyanotic heart diseases, it is mandatory to keep Botallo's ductus arteriosus, a fetal bypass of the lung, patent. For that purpose, a prolonged administration of prostaglandin E1 is the rule. As one of its complications abnormal cortical hypertrophy of the diaphysis of long bones (hypertrophic osteoarthropathy) takes place while the change in the epiphysis and metaphysis are much less prominent (Tomiie *et al.* 1982). In this case, HAC itself could have protected the joints, but hyperemic effects were confined to the diaphysis over which there were non anatomical or functional capacitors and the systemic blood flowed into the bones directly. This is a "no HAC" example. Thus, this clinical experience shows the evolutionary mechanism of the arterial circles.

Expansion of Hunter's arterial capacitor function. The original Hunter's principium of arterial capacitor function can be applied to other organ systems such as the brain and heart (Figure 4). This expansion is not necessarily by homology but by analogy. Graph A schematizes the original HAC. Graph B is for the cerebral circulation: the arterial branches from Willis' arterial circle (WAC) are thicker than in the case of prototype HAC.

WAC locates at the base of the brain. The human brain is characterized by its prompt activation to any stimulations. Resultantly, the velocity of the middle cerebral artery (MCA) is made greater than the original carotid blood flow so as to ensure instantaneous perfusion on demand (Suwa *et al.* 1969): (i) The blood flow rate in the brain is increased to 6.4×10^{-9} ml per sec as compared with 4.4×10^{-9} for the kidney; (ii) The arterial branches are made thicker and longer than in the case of the renal arteries so as to minimize the fall of blood pressure; and thus, (iii) the greater blood flow can be conveyed instantly. Then WAC's capacitor function is possible so as not to interfere with that kind of cerebral circulation.

Graph C stands for the coronary circulation: There is no anatomical arterial circulus there. However, we may imagine a functional arterial circle in diastole when the aortic valves close because of the aortic back pressure and the blood flows into the myocardium through the coronary sinus (CAC). As the coronary arteries penetrate the myocardium from the outside into the inside, the myocardial perfusion takes place during the diastole for the inner myocardium and, that for the outer portions, during systole by forced diffusion.

We may well subcategorize WAC and CAC into pre-WAC, WAC and post-WAC, and pre-CAC, CAC and post-CAC for descriptive purposes.

Cerebral vascular diseases. The cerebral vascular accidents (CVAs) of cerebral hemorrhage from ruptured microaneurysms (MA) and infarction as the result of embolization of atherosclerotic plaques (AS) are post-WAC events. CVAs may represent an evolutionary cost of bipedalism in man because of the largeness of the brain.

How does the cerebral circulation function in response to bipedality? The vasculature has to be so designed as to increase the central blood supply to sites of the brain of activity on the instant of demand as follows (Fig. 4B): (i) ICA (internal carotid artery) is placed at right angle to WAC; (ii) ICA's blood flow is biased almost directly to MCA; and (iii) acute perfusion of those brain centers of activity occurs on demand only. (iv) Then, the WAC capacitance is really not so great because slight blood pressure changes in the basilar or vertebral arteries are sharply echoed by the development of vertigo. Therefore, (v) Suwa's theorem of cerebral vasculature is critically important: (a) to increase the blood flow rate; (b) to lengthen the arterial branches of larger diameters, and (c) to minimize blood pressure fall (Suwa *et al.* 1963). The transcranial Doppler data on blood flow velocities are in agreement with the theorem (Ringelstein *et al.* 1990).

Then, what is the reality of CVAs? (1) 70% of cerebral hemorrhage in MCA (the putamen and thalamus); (2) MAs are responsible for hemorrhage; (3) Because of the short distance of MCA from the internal carotid artery (ICA), the impacts of the ICA's arterial blood flow can be not sufficiently attenuated; (4) Thus, the expected capacitor function of has WAC has been greatly spoiled especially when hypertension impinges upon the vasculature.

Atherosclerosis in the proximal internal carotid artery is usually severe in the first 2 cm and arises in the posterior walls (Kistler *et al.* 1969). Then, the aortic ejecta are twisted and exert shearing forces upon the posterior wall of the internal carotid artery at its origin as well as cranialward impact, and the resultant damage will entice atherosclerotic plaques. When an embolus reaches WAC and passes it to embolize in the cerebral arterial branches, embolization and cerebral infarction ensues. Thus, a pre-WAC pathology harms the post-WAC structures.

MA in the post-WAC regions is another sequela of intimal damage propagated and perpetuated by the reactions relative to superoxide radical toxicity. Superoxide anions that mar the arterioles are produced because of aging-related uncoupling of blood flow and oxygen consumption, especially in the cerebral basal ganglial territory (Yamaguchi *et al.* 1986) Once should it take place, blood torrents as predicted by Suwa's theorem are also to amplify the damage to the formation of microaneurysms, for there is no effective device yet to counteract the adverse active oxygen actions there. Then, this is post-WAC event and is one of the evolutionary *loci minores resistentiae.*

Myocardial infarction. Myocardial infarction is a necrotic sequela of coronary artery occlusions.

In the heart, there is no anatomical HAC analogue but a functional one, say, "coronary arterial circle (CAC)". It is formed during diastole with the column of blood as soon as the peripheral resistance to the cardiac outflow or the back pressure establishes itself and the aortic valves close themselves (Fig. 4C). It has an ample capacitance, and provides steady coronary blood flow at reduced blood pressure, pulse pressure and velocity through the coronary arteries. It thus minimizes arterial impacts onto the vascular walls. The coronary arteries, the post-CAC arteries, as they resemble MCA, follow Suwa's theorem, too, so as to promptly convey blood to sites of demand. One of the peculiarities of the post-CAC circulation is that the inner myocardium is nourished directly by the coronary circulation, and the blood supply to the outer myocardium takes place through diffusion by dint of myocardial contraction during systole (Hoffman *et al.* 1985). This CAC - post-CAC balance will be critically impaired through two mechanisms: (1) All those pre-CAC risk factors such as hypertension, hyperlipidemia, glycation from hyperglycemia and cigarette smoking that will predispose to coronary arteriosclerosis; and (2) post-CAC arteriosclerotic progression. The primary predisposition of arteriosclerosis is the proximal coronary arteries presumably because of the above-mentioned mechanical impact shearing twists upon the arterial walls during cardiac activity. In addition, vibration at sites of branching emerging with uneven branching as in air lift (Alonso and Finn 1970) will foster formation, dislodgement an further embolic metastasis of arteriosclerotic plaques. The above-mentioned pre-CAC risk factors will also be counted here.

CVA sequlae as the disorders of hominized cognition. Asymmetry is human. Asymmetry is true of the human brain. The left brain fulfills analytical jobs while the right hemisphere, holistic. In the left hemisphere, there are centers for speech. In the right, centers for spatial recognition and music. This asymmetric brings forth misery to men as soon as portions of the hemispheric cortices are destroyed in one way or another. It is a high contrast from other primate monkeys in which symmetry of the hemispheres, both morphological and functional, is strictly observed. thus, while the asymmetry of the brain has expanded human capability ion one hand, the same hominization imposes deadly tolls on the other. Up to the higher primates, the right and left cerebral hemispheres function identically. However, with mankind, the right and left hemispheres function differently. In about 95% of people, the centers for sensory and motor speech are located in the left hemisphere whereas those activities of pattern analysis, music and spatial orientation are located in the right hemisphere (Eccles 1989). This is the hominization of the brain. This cerebral asymmetry can torment the human being by depriving of the ability to compensate the function of one hemisphere with that of the counterpart.

An operated breast cancer patient developed signs and symptoms of spatial disorientation. Cerebral infarction limited to the right parieto-occipital cortex (Fig. 9, left) produced difficulty in spatial recognition, and she used to fail to return to her ward after taking a walk in the corridor. A cerebral perfusion study of her showed identical areas of reduced blood flow (Fig. 9, right).

In this way, evolutionary concepts will help us decipher disease mechanisms and hopefully help in designing treatment of certain disorders.

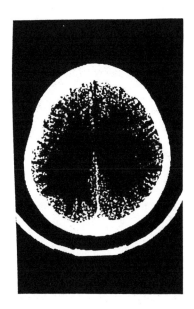 

FIGURE 9. **Cost of hominization in the brain: Asymmetric use of the human brain in hominization.** It has greatly expanded the human's intellectual capability, but it has also broken the compensatory function of the symmetric brain. A cerebral infarction in an area of the right hemisphere responsible for spatial cognition disabled this 70-year-old woman by depriving adequate orientation. She used to fail to return to her after walking in the corridor. Left: An XCT of the brain. The low density areas represented those of hypoperfusion (Right).

References

[1] Alonso, M. & Finn, E.J. (1970) *Physics*, Addison-Wesley, Reading, Mass.

[2] Bennett, G.A. (1957) The bones. In: *Pathology*, 3rd ed., WAD Anderson, ed. Mosby, St. Louis. pp. 1189–1250.

[3] Brookes, M. (1971) *The Blood Supply of Bone. An Approach to Bone Biology*, Butterworths, London.

[4] Eccles, J.C. (1989) *Evolution of the Brain: Creation of the Self*, Routledge, London.

[5] Fujita, S. (1988) Evolution of the brain. In: *Saishin Noh no Kagku* (Frontier of Brain Science). Vol I. Basic Aspects, M. Ito & T. Kuwabara, eds., Dobun Shoin, Tokyo. pp. 29–57.

[6] Greenwald A.S. (1991) Biomechanics of the hip. In: *The Hip and Its Disorders*, M.E. Steinberg, ed., Saunders, Philadelphia. pp. 47–55.

[7] Harty M. (1991) Anatomy. In: *The Hip and Its Disorders*, M.E. Steinberg, ed., Saunders, Philadelphia. pp. 27–46.

[8] Hodgkin T. (1832) On some morbid appearances of the absorbent glands and spleen. *Med-Chir. Transactions* **17**: 68–114.

[9] Hoffman, J.I.E., Baer, R.W., Hanley F.L. & Messive, L.M. (1985) Regulation of transmural myocardial blood flow. *Trans. ASME* **107**:2–9.

[10] Horowitz, N.H. (1945) On the evolution of biochemical synthesis. *Proc. Natl. Acad. Sci., USA* **31**: 153–155.

[11] Ishizaka K. & Ishizaka T. (1966) Physicochemical properties of reaginic antibody. I. Association of reaginic activity with immunoglobulin other than γA or γG-globulin. *J Allerg.* **37**: 169.

[12] Kistler, J.P., Ropper, A.A., Martin, J.B. (1991) Cerebrovascular diseases. In: *Harrison's Principles of Internal Medicine*, 12th ed., J.D. Wilson, E. Braunwald, K.J. Isselberger, R.G. Petersdorf, J.B. Martin, A.S. Fauci & R.K. Root, McGraw-Hill, New York. pp. 1977–2002.

[13] Leader, W.G. (1969) The Chediak-Higashi anomaly... an evolutionary concept of disease. In: *National Cancer Institute Monograph 32. Comparative Morphology of Hematopoietic Neoplasms*, National Cancer Institute, Bethesda, pp. 337–339.

[14] Margulis, L. (1981) *Symbiosis and Cell Evolution*, Freeman, San Francisco.

[15] Nakamura, H. (1982) *Saibu no Kigenn to Shinka* (The Origin of the Cell and Its Evolution), Baifukan, Tokyo. (in Japanese)

[16] Okuyama, S. & Mishina, H. (1984) Evolutionary concepts of cancer and its evidence observable in the radiological sciences. *Eizo Joho Medical* **16**: 865–871. (in Japanese)

[17] Okuyama, S. & Mishina, H. (1986) Hypothesis: Evolutionary concepts of cancer, 45th Ann. Gen. Assemb. Cancer, Sapporo, Oct. 22, 1986.

[18] Okuyama, S. & Mishina, H. (1987) Fanconi's anemia as a Nature's evolutionary experiment on carcinogenesis. *Tohoku J. Exp. Med.* **153**: 87–102.

[19] Okuyama, S. & Mishina, H. (1988) Cancer incidence in the population of Nagasaki City 30 years after atomic bombing. *Tohoku J. Exp. Med.* **155**:23-39.

[20] Okuyama, S. & Mishina, H. (1990) *Evolution of Cancer*, University of Tokyo Press, Tokyo.

[21] Okuyama, S. (1991) Probable binary fission in a Reed-Sternberg cell. *Tohoku J. Exp. Med.* **164**: 247–249.

[22] Okuyama, S. (1992) Impacts of the evolutionary concepts of cancer on the study of human diseases. *Tohoku J. Exp. Med.* **168**: 445–448.

[23] Okuyama, S. (1996) Evolution of Hodgkin's disease. *Med. Hypotheses* **47**: 199–213.

[24] Okuyama, S. & Sato, A. (1996) Are the cerebral basal ganglia responsible for the development of microaspiration pneumonia? In: *Proceedings of the Second Tohoku Conference on Cerebral Circulation*, Sendai, September 30, 1995. pp. 43–45. (in Japanese)

[25] Okuyama, S. (1997a) Hominization of Cognition, Hominization of Computer. In: *Proceedings of the Second International Conference on Cognitive Technology*, Aizu, August 25-28, 1997, IEEE Computer Society Press.

[26] Okuyama, S. (1997b) The first attempt at radioisotopic evaluation of the integrity of the nose-brain barrier. *Life Sci.* **60**: 1881–1884.

[27] Ringelstein, E.B., Kahlscheuer G., Niggemeyer E. & Otis, S.M. (1990) Transcranial Doppler sonography. Anatomical landmarks and normal velocity values. *Ultrasound Med. Biol.* **16**: 745–761.

[28] Sasaki, T., Yamamoto, M. & Sakka, M. (1981) Implications of thymiadine labeling index in the growth kinetics of human solid tumors. *Gann* **72**: 181–188.

[29] Setala, K. (1984) Carcinogenesis — Devolution towards an ancient nucleated pre-eukaryotic level. *Med. Hypotheses* **15**: 209–230.

[30] Sevitt, S. & Thompson, R.G. (1965) The distribution and anastomoses of arteries supplying the head and neck of the femur. *J. Bone Joint Surg. (Br)* **47**: 560–573.

[31] Shannon, R.P., Mahler, K.A., Santianga, J.T., Royal, H.D. & Wei, J.Y. (1992) Comparison of differences in hemodynamic response to passive postural stress in healthy subjects 70 years and 30 years years of age. *Am. J. Cardiol.* **67**: 1110–1116.

[32] Snell, O. (1982) Die Abhaengigkeit des Hirngewichtes von der Koerpergewichte und den Geistesfaehigkeiten. *Arch. Psych.* **23**: 436–446.

[33] Stobbe, H. (1983) Zur Zytogenese und Zytomorphologie der Hodgkin- und Sternberg-Reed-Zellen. *Arch. Geschwulstforsch.* **53**: 115–123.

[34] Sugano, H. (1980) Natural history of human cancers. *Tr. Soc. Pathol. Jpn.* **69**: 27–57. (in Japanese)

[35] Suwa, N., Niwa, T., Fukasawa, H. & Sasaki, Y. (1963) Estimation of intravascular blood pressure gradient by mathematical analysis of arterial casts. *Tohoku J. Exp. Med.* **79**: 168–198.

[36] Tomiie, F., Yamamoto, A., Murakami, K. & Hamaoka, K. (1982) Bone changes in infants induced by prostaglandin E1 as secondary hypertrophic osteoarthropathy. *Nippon Acta Radiol.* **42**: 1127–1136. (in Japanese with English summary)

[37] Wintrobe, M.M., Lee, G.R., Boggs, D.R., Bithell, T.C., Foerster, J., Athens, J.W., Lukens, J.N. (1981) *Clinical Hematology*, 8th ed., Lea & Febiger, Philadelphia.

[38] Yamaguchi, T., Kanno, I., Uemura, K., Shishido, F., Inugami, A., Ogawa, T., Murakami, M. & Suzuki, K. (1986) Reduction in regional cerebral metabolic rate of oxygen during aging. *Stroke* **17**: 1220–1228.

DEPARTMENT OF RADIOLOGY, TOHOKU ROSAI HOSPITAL, SENDAI 981, JAPAN

Lectures on Mathematics in the Life Sciences
Volume **26**, 1999

Maximal Circular Codes and Applications to Theoretical Biology

Giuseppe Pirillo

ABSTRACT. Interesting maximal circular codes containing 20 trinucleotides have been recently discovered by the team of theoretical biology of Didier Arquès of University of Marne-la-Vallée (France). A possible biological application of this result is the automatic finding of genes in genomes.

1. Introduction.

Hereafter we give some information on a recent result (more details can be found in [1] and [2]): following some research efforts and using a particularly simple statistical investigation, the 64 trinucleotides have been partitioned in 3 classes which have several surprising properties. For example, after cancellation of the trinucleotides of the form XXX, with $X \in \{A, C, G, T\}$, each of them is a maximal circular code.

We end this note with a short information on a work in progress. We are working on a method that, using the above-mentioned results of [1] and [2], enables us to determine automatically genes in genomes.

2. Some notations of theoretical computer science.

The terminology concerning sequences (words) is that of [4]. In any case, for the aim of this note we introduce some little modifications in accord with the "correspondence between the computer theory of languages and the gene structure" used in several papers of the team of theoretical biology of Didier Arquès of University of Marne-la-Vallée.

Given an *alphabet B*, the *free monoid* (resp. *free semigroup*) over B is denoted by B^* (resp. B^+). An element of B is a *letter* (*nucleotide* or *base*). We use a four letter alphabet $B = \{A, C, G, T\}$ where A, C, G, T are respectively the bases *Adenine, Cytosine, Guanine, Thymine*. An element of B^* is a *sequence* (*word*) and, in particular, a sequence of length 3 is a *trinucleotide*. The *empty sequence* is denoted by 1. A subset of B^* is, by definition, a *language* (or *gene population*).

1991 *Mathematics Subject Classification*. Primary 68Q45, 92D20 and Secondary 05E20, 20M35, 92B05.

M. P. Schützenberger has been the main architect of the theory of (variable length) codes (see [3]). Some notions of this theory, which is in strong relations with physics and biology (for instance), are explicitly defined hereafter.

Definition. *A language X in B^+ is a code if for each $n, m \geq 1$ and for each $x_1, \cdots, x_n, x'_1, \cdots, x'_m$ in X the condition*

$$x_1 \cdots x_n = x'_1 \cdots x'_m$$

implies $n = m$ and, for $i = 1, \cdots, n$,

$$x_i = x'_i.$$

Definition. *A language X in B^+ is a circular code if, for each $n, m \geq 1$ and for each $x_1, \cdots, x_n, x'_1, \cdots, x'_m$ in X, $p \in B^*$ and $s \in B^+$, the conditions*

$$sx_2 \cdots x_n p = x'_1 \cdots x'_m$$

and

$$x_1 = ps$$

imply $n = m$, $p = 1$ and, for $i = 1, \cdots, n$,

$$x_i = x'_i.$$

In order to have an intuitive meaning of the two previous definitions, one can imagine writing on a straight line when we speak of codes and writing on a circle when we speak of circular codes; in both cases unique decipherability is required. Note that the condition of being a circular code is much stronger than that of being a code.

X is a *maximal* circular code means that X is a circular code such that if X is contained in another one Y, then $X = Y$.

3. Three maximal circular codes.

Let $w = w(0)w(1)\ldots w(n)$ be a sequence and let $i \in \{0, 1, 2\}$. We say that the sequence $l_1 l_2 l_3$ has an occurrence in the class i if there exists j smaller than or equal to $n - 2$ such that $j = i \bmod 3$ and $w(j)w(j + 1)w(j + 2) = l_1 l_2 l_3$.

For example, if $w = ACAACAGTCTCTCACACA$ then ACA has three occurrences in class 0, one in class 1 and no occurrence in class 2. In this example, clearly ACA prefers the class 0.

Large gene populations encoding proteins of eukaryotes or prokaryotes taken from EMBL Nucleotide Sequence Data Library have been considered in order to associate to each trinucleotide the class containing it most frequently.

Two independent statistical investigations (one for eukaryotes and the other for prokaryotes) have given results which agree. The following table shows the partition of the 64 trinucleotides in T_0 (preferred class 0), T_1 (preferred class 1) and T_2 (preferred class 2):

$T_0 = \{AAA, AAC, AAT, ACC, ATC, ATT, CAG, CTC, CTG, GAA, GAC$
$GAG, GAT, GCC, GGC, GGT, GTA, GTC, GTT, TAC, TTC, TTT\}$

$T_1 = \{AAG, ACA, ACG, ACT, AGC, AGG, ATA, ATG, CCA, CCC, CCG$
$GCG, GTG, TAG, TCA, TCC, TCG, TCT, TGC, TTA, TTG\}$

$T_2 = \{AGA, AGT, CAA, CAC, CAT, CCT, CGA, CGC, CGG, CGT, CTA$
$CTT, GCA, GCT, GGA, GGG, TAA, TAT, TGA, TGG, TGT\}.$

Put $X_0 = T_0 - \{AAA, TTT\}, X_1 = T_1 - \{CCC\}, X_2 = T_2 - \{GGG\}$. Each
of the sets X_0, X_1, X_2 contains 20 trinucleotides and has the following remarkable
properties.

Property of complementarity.

As the double helix of DNA consists of two sequences of nucleotides s_1 and
s_2 oriented in opposite sense and as to each occurrence of A (resp. C, G, T) in s_1
corresponds in s_2 an occurrence of T (resp. G, C, A) [5], we say that the complement
of the trinucleotide $l_1 l_2 l_3$ is the trinucleotide $c(l_1 l_2 l_3) = c(l_3)c(l_2)c(l_1)$ where, for
each i in $\{0, 1, 2\}$, $c(l_i) = T$ (resp. A, G, C) if $l_i = A$ (resp. T, C, G). The property
of complementarity (easy to verify) is the following:

$$c(X_0) = X_0, \quad c(X_1) = X_2, \quad c(X_2) = X_1$$

i.e., in other words, X_0 is self-complementary and, on the other hand, X_1 and X_2
are complementary to each other.

Property of circularity.

If $l_1 l_2 l_3$ is a trinucleotide, we pose $p(l_1 l_2 l_3) = l_2 l_3 l_1$. The property of circularity
(easy to verify too) is the following:

$$p(X_0) = X_1, \quad p(X_1) = X_2, \quad p(X_2) = X_0.$$

Maximal circular code.

Theorem [1, 2]. *The sets X_0, X_1, X_2 are maximal circular codes.*

This property of X_0, X_1, X_2 is the most difficult to prove. The argument is
based on the properties of the so-called *flower automaton* (see [1] and [2] for details).

4. An application: the automatic research of genes in genome.

The number of cases to examine in order to find maximal circular codes on a
four letter alphabet is $3^{20} = 3,486,784,401$. The number of maximal circular codes
on a four letter alphabet is 12,964,440, from which only 216 have properties similar
to that of code X_0. So the probability to find it was really very small.

Didier Arquès and the author of this note are working on the following possible
application of the above mentioned results of [1] and [2]: to find genes automatically
in a genome. To this aim we are preparing an appropriate program, which is based
on the frequencies of the codes X_0, X_1, X_2 in the genes and which will be applied
to the now available complete genomes.

Acknowledgement

The author thanks Didier Arquès for his help in the preparation of this note.

References

[1]. D. G. Arquès and C. J. Michel, A possible code in the genetic code. In: Ernst W. Mayr Claude Puech, eds. STACS 95 - *12th Annual Symposium on Theoretical Aspects of Computer Science, Munich, Germany, March 2 - 4, 1995 Proceedings*, Lecture Notes in Computer Science, Springer Verlag, Vol. 900, (1995) 640-651.

[2]. D. G. Arquès and C. J. Michel, A complementary circular code in the protein coding genes, *Journal of Theoretical Biology*, **182** (1996) 45–58.

[3]. J. Berstel and D. Perrin, *Theory of Codes*, Academic Press, London, 1985.

[4]. M. Lothaire, *Combinatorics on Words*, Addison-Wesley, London, 1983.

[5]. J. D. Watson and F. H. C. Crick, A structure for deoxyribose nucleic acid, *Nature* **171**, (1953) 737-738.

IAMI CNR, Viale Morgagni 67/A, 50134 Firenze, Italy, and, Institut Gaspard Monge, Bâtiment IFI, Université de Marne-la-Vallée, 2 rue de la Butte Verte, 93160 Noisy-le-Grand, France

E-mail address: pirillo@udini.math.unifi.it

Lectures on Mathematics in the Life Sciences
Volume **26**, 1999

Sorting Permutations and Its Application in Genome Analysis

Qian-Ping Gu, Shietung Peng, and Qi-Ming Chen

ABSTRACT. An efficient approach for checking the similarity between genomes on a large scale is to compare the order of appearance of identical genes in the two species. The two gene sequences differ not because of local mutations, but because of global rearrangements such as reversals, transpositions, and so on. Given the sequences of the identical genes of two species, if we express one sequence by $I = (12...n)$ then the other sequence can be expressed by a permutation $\pi = (\pi_1 \pi_2 ... \pi_n)$ of $\{1, 2, ..., n\}$. Checking the similarity between genomes based on global rearrangements leads to a combinatorial problem of finding a shortest series of rearrangements that sorts the permutation π into the identity I. In this paper, we propose algorithms for permutation sorting by reversals and transpositions.

1. Introduction

With the fast progress of the Human Genome Project, genetic and DNA data is accumulating rapidly. Many studies have shown that one of the efficient approaches for checking the similarity between genomes on a large scale is to compare the order of appearance of identical genes in the two species. In this approach, the similarity can be measured by the distance between two species. The data is the order in which the gene loci (rather than the nucleotides or amino acids) lie on the chromosome. The sequences of loci for two species differ not because of insertions, deletions, or mutations, but because of global rearrangements such as reversals, transpositions, and so on. Given the sequences of the identical genes of two species, if we express one sequence by $I = (12...n)$ then the other sequence can be expressed by a permutation $\pi = (\pi_1 \pi_2 ... \pi_n)$ of $\{1, 2, ..., n\}$. Checking the similarity between genomes based on global rearrangements leads to a combinatorial problem of finding a shortest series of rearrangements that sorts the permutation π into the identity I.

Mathematical analysis of genome rearrangement problems was initiated by Sankoff [**SCA90, San92**]. Approximation algorithms for sorting permutations by reversals were proposed by Kececioglu and Sankoff [**KS93**], and then Bafna and Pevzner (with an improved error bound) [**BP96**]. Recently, Caprara proved that sorting permutations by reversals is NP-hard [**Cap97**]. An exact polynomial time algorithm for sorting *signed permutation* (a permutation π on $\{1, 2, ..., n\}$ with + or − sign associated with every element π_i of π) by reversals was given by Hannenhalli and Pevzner [**HP95**]. Kaplan et al. [**KST97**] simplified the algorithm of Hannehalli and Pevzner. Bafna and Pevzner studied the problem of

---

1991 *Mathematics Subject Classification*. Primary 05A05, 68P10, 92D99; Secondary 68R05.

sorting permutations by transpositions. They gave the first constant factor polynomial time approximation algorithm for the problem [**BP95**]. Their algorithm is a 1.5-approximation one.[1] Gu, Peng, and Sudborough studied the problem of sorting signed permutation using reversals, transpositions, and reversal+transpositions simultaneously. They gave a 2-approximation algorithm for the problem [**GPS96**].

In general, it is computationally difficult to find the minimum number of operations (reversals/transpositions) for sorting permutations. Developing heuristic and approximation algorithms for sorting permutations has been a key in the analysis of genome rearrangements. In this paper, we propose such algorithms. Especially, we give a 1.5-approximation algorithm for sorting permutations by transpositions only. Although our algorithm has the same error bound as that of [**BP95**], it is much simpler than the previous one. We also give a heuristic algorithm for sorting signed permutations by reversals, transpositions, and reversal+transpositions. The experiment shows that the heuristic algorithm sorts randomly generated permutations in an almost optimal number of rearrangements.

In the next section, we give the preliminaries of the paper. Sorting permutations by transpositions is discussed in Section 3. Section 4 gives the heuristic algorithm, and the final section concludes the paper.

2. Preliminaries

Let $\pi = (\pi_1 \pi_2 ... \pi_n)$ be a permutation of $\{1, 2, ..., n\}$. For $1 \leq i < j \leq n + 1$, a *reversal* $r(i, j)$ is the permutation

$$\left(\begin{array}{l} 1...i - 1\ \mathbf{i}\ \mathbf{i+1}...\mathbf{j} - \mathbf{1}\ j...n \\ 1...i - 1\ \mathbf{j} - \mathbf{1}...\mathbf{i} + \mathbf{1}\ \mathbf{i}\ j...n \end{array} \right).$$

$\pi \cdot r(i, j) = \pi_1...\pi_{i-1}\pi_{j-1}...\pi_{i+1}\pi_i\pi_j...\pi_n$, i.e., $\pi \cdot r(i, j)$ has the effect of reversing the order of $\pi_i, \pi_{i+1}, ..., \pi_{j-1}$. For $1 \leq i < j \leq n + 1$ and $1 \leq k \leq n + 1$ with $k \notin [i, j]$, a *transposition* $t(i, j, k)$ is the permutation

$$\left(\begin{array}{l} 1...i - 1\ \mathbf{i}\ \mathbf{i+1}...\mathbf{j} - \mathbf{1}\ j...k - 1\ k...n \\ 1...i - 1\ j...k - 1\ \mathbf{i}\ \mathbf{i+1}...\mathbf{j} - \mathbf{1}\ k...n \end{array} \right).$$

$\pi \cdot t(i, j, k) = \pi_1...\pi_{i-1}\pi_j...\pi_{k-1}\pi_i...\pi_{j-1}\pi_k...\pi_n$, i.e., $\pi \cdot t(i, j, k)$ has the effect of moving $\pi_i\pi_{i+1}...\pi_{j-1}$ to a new location of π between π_{k-1} and π_k. For $1 \leq i < j \leq n + 1$ and $1 \leq k \leq n + 1$ with $k \notin [i, j]$, a *reversal+transposition* $rt(i, j, k)$ is the permutation

$$\left(\begin{array}{l} 1...i - 1\ \mathbf{i}\ \mathbf{i+1}...\mathbf{j} - \mathbf{1}\ j...k - 1\ k...n \\ 1...i - 1\ j...k - 1\ \mathbf{j} - \mathbf{1}...\mathbf{i} + \mathbf{1}\ \mathbf{i}\ k...n \end{array} \right).$$

$\pi \cdot rt(i, j, t) = \pi_1...\pi_{i-1}\pi_j...\pi_{k-1}\pi_{j-1}...\pi_{i+1}\pi_i\pi_k...\pi_n$, i.e., $\pi \cdot rt(i, j, k)$ has the effect of reversing $\pi_i\pi_{i+1}...\pi_{j-1}$ and then moving $\pi_{j-1}...\pi_i$ to a new location of π between π_{k-1} and π_k. We will call the reversal, transposition, and reversal+transposition *operations*.

EXAMPLE 2.1. Let $\pi = (14352)$. Then $\pi \cdot r(1, 4) = (34152), \pi \cdot t(1, 4, 5) = (51432)$, and $\pi \cdot rt(1, 4, 5) = (53412)$.

The distance between two permutations π and σ is the minimum number of operations $\rho_1, ..., \rho_t$ such that $\pi \cdot \rho_1 \cdot \rho_2 \cdots \rho_t = \sigma$. Note that the distance between π and σ equals that between $\sigma^{-1}\pi$ and the identity $I = (12...n)$. Thus, we only concentrate on finding the distance $d(\pi)$ between π and I.

---

[1] For a permutation π, let $d(\pi)$ be the minimum number of transpositions to sort π into I. An α-approximation algorithm for sorting a permutation is an algorithm which, given any π, finds a series of transpositions $\rho_1, ..., \rho_t$ such that $\rho_1, ..., \rho_t$ sort π into I and t satisfies $d(\pi) \leq t \leq \alpha d(\pi)$.

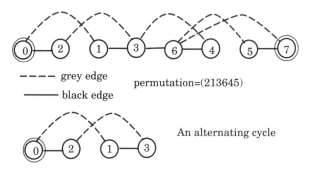

FIGURE 1. The breakpoint graph $G(\pi)$ of $\pi = (213645)$.

A *signed permutation* is a permutation π on $\{1, 2, ..., n\}$ with $+$ or $-$ sign associated with every element π_i of π. For example, $(+1 - 5 + 4 - 3 + 2)$ is a signed permutation. The identity $I = (+1 + 2... + n)$. A reversal $r(i, j)$ on a signed permutation changes both the order and the signs of the elements within the fragment $\pi_i \pi_{i+1} ... \pi_{j-1}$. For signed permutation π, we are interested in finding the minimum number of operations to sort π into the identity $(+1 + 2... + n)$.

Bafna and Pevzner introduced the notion of *breakpoint graph* in their study for sorting by reversals only [**BP96, HP95**]. Since our argument is also based on the *breakpoint graph*, we introduce it in some detail here.

Let π be an arbitrary permutation. Extend $\pi = \pi_1 \pi_2 ... \pi_n$ by adding $\pi_0 = 0$ and $\pi_{n+1} = n + 1$. Let $i \sim j$ if $|i - j| = 1$. We call a pair of consecutive elements π_i and π_{i+1} an *adjacency* if $\pi_i \sim \pi_{i+1}$, otherwise a *breakpoint*. Define a *breakpoint graph* $G(\pi)$ of π as follows: There are $n + 2$ nodes $0, 1, 2, ..., n, n+1$ in $G(\pi)$. There is a grey edge between i and j if $i \sim j$ and i, j are not consecutive in π. There is a black edge between i and j if (i, j) is a breakpoint. The graph $G(\pi)$ for $\pi = (213645)$ is given in Figure 1. Notice that the number of grey edges equals the number of black edges in $G(\pi)$, and equals to the number of breakpoints in π. The *breakpoint* graph $G(I)$ for the identity I has no edge.

Define a transformation from a signed permutation π of n elements to an unsigned permutation π' of $2n$ elements as follows: replace $+i$ with $(2i - 1, 2i)$ and replace $-i$ with $(2i, 2i - 1)$ for $1 \le i \le n$. Notice that the identity $I = (+1 + 2... + n)$ is transformed into the unsigned identity $I' = (1234...(2n - 1)2n)$. Given any sequence of operations $\rho_1, ..., \rho_t$ which transforms π into σ, obviously, there is a sequence $\rho'_1, ..., \rho'_t$ which transforms π' into σ'. On the other hand, for any sequence of operations $\rho'_1, ..., \rho'_t$ transforming π' into σ' such that no operation breaks any pair of $(2i - 1, 2i)$ or $(2i, 2i - 1)$ for $1 \le i \le n$, then there is a sequence of operations $\rho_1, ..., \rho_t$ that transforms π into σ. In what follows, we assume that any operation on the transformed unsigned permutation never breaks any pair of $(2i - 1, 2i)$ or $(2i, 2i - 1)$. Based on this assumption, the signed permutation π and the transformed unsigned permutation π' are equivalent for our purpose. When we refer to the breakpoint graph of a signed permutation, it is implied that we refer to the breakpoint graph of the transformed unsigned permutation. Figure 2 gives the breakpoint graph of $G(\pi)$ for $\pi = (+1 + 5 + 4 + 3 + 2)$.

A sequence of nodes $v_1, v_2, ..., v_m = v_1$ is called a *cycle* in a graph G if $(v_i, v_{i+1}) \in E(G)$ for $1 \le i \le m - 1$. A cycle in a breakpoint graph $G(\pi)$ is called *alternating* if the colors of every two consecutive edges of this cycle are distinct. We define the length of an alternating cycle the number of black edges (breakpoints) in the cycle. In what follows, we

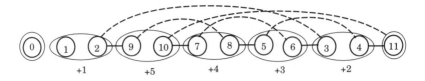

FIGURE 2. The breakpoint graph $G(\pi)$ of $\pi = (+1 + 5 + 4 + 3 + 2)$.

use cycle for alternating cycle unless otherwise stated. For example, cycle $0, 2, 3, 1, 0$ of the graph $G(\pi)$ in Figure 1 is alternating and has length 2.

3. Sorting permutations by transpositions

Given a permutation π, decompose the breakpoint graph $G(\pi)$ into edge-disjoint cycles, and then remove the cycles from $G(\pi)$ was a principle approach of sorting permutations in the previous study. However, the decomposition of $G(\pi)$ may not be unique if π is unsigned. This has been a bottleneck in the above approach. Bafna and Pevzner proposed an alternative graph model *edge-colored cycle graph* to get an unique cycle decomposition and gave a 1.5-approximation algorithm for sorting permutations by transpositions [**BP95**]. The previous studies also revealed that the long cycles in the breakpoint graphs or edge-colored cycle graphs make the analysis of sorting permutations difficult [**BP95, HP95**].

Now, we give an alternative proof of the existence of a 1.5-approximation algorithm for sorting permutations by transpositions. We apply breakpoint graphs to our analysis instead of edge-colored cycle graphs based on the observation that a transposition ρ never changes the sign of any element π_i when ρ is applied to a signed permutation π. Therefore, a permutation $(\pi_1...\pi_n)$ can always be regarded as a signed permutation $(+\pi_1...+\pi_n)$. The breakpoint graph $G((+\pi_1... + \pi_n))$ has an unique cycle decomposition. (Figure 2 gives the breakpoint graph of $G(\pi)$ for $\pi = (15432)$.) To simplify the analysis for long cycles in the breakpoint graphs, we cut the long cycles into short ones by a *padding* technique developed by Hannenhalli and Pevzner [**HP95**]. Our proof is much simpler than that of Bafna and Pevzner. The simplified proof also leads to a 1.5-approximation algorithm which is simpler than the previous algorithm of Bafna and Pevzner. Also the time complexity of our algorithm is expected to be more efficient than that of Banfa and Pevzner for practical data.

Given a permutation π, an operation ρ, and a permutation $\pi' = \pi \cdot \rho$, the numbers of breakpoints and the numbers of cycles in $G(\pi)$ and $G(\pi')$ have been used to measure the efficiency of the operation ρ. In particular, let $b(\pi)$ be the number of the breakpoints and $c(\pi)$ be the number of cycles with odd length in the breakpoint graph $G(\pi)$ of π. It has been known that for $\pi' = \pi \cdot \rho$, $(b(\pi) - b(\pi')) + (c(\pi') - c(\pi)) \leq 2$ [**GPS96**]. This provides a lower bound $(b(\pi) - c(\pi))/2$ on sorting signed permutations by any operation of reversals, transpositions, and reversals+transpositions.

Call an operation ρ an i-move on π if for $\pi' = \pi \cdot \rho$, $b(\pi) - b(\pi') + c(\pi') - c(\pi) = i$. From the lower bound given above, a transposition is an i-move with $i \leq 2$.

Call a cycle a k-cycle if its length is k. Call a k-cycle a long cycle if $k > 3$ otherwise a short cycle. Notice that a short cycle is either a 2-cycle or a 3-cycle. Now, we construct a 1.5-approximation algorithm for sorting permutations by transpositions. Our construction starts from cutting long cycles in $G(\pi)$ into short ones by a *padding* technique introduced by Hannenhalli and Pevzner [**HP95**].

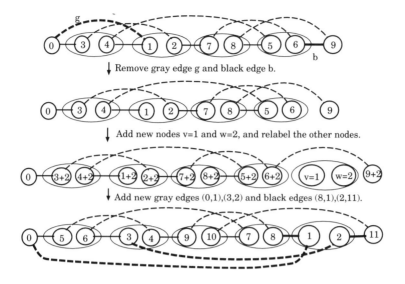

FIGURE 3. An example of cutting one cycle into two by padding.

For a permutation $\pi = (\pi_1...\pi_i\pi_{i+1}...\pi_n)$ of $\{1, 2, ..., n\}$, let $b = (\pi_i, \pi_{i+1})$ be a black edge and $g = (\pi_r, \pi_s)$ be a gray edge belonging to a cycle C in $G(\pi)$. A (g, b)-*padding* of π is a permutation $\hat{\pi} = (\hat{\pi}_1...\hat{\pi}_i vw\hat{\pi}_{i+1}...\hat{\pi}_n)$ of $\{1, 2, ..., n+2\}$, where $v = \min\{\pi_r, \pi_s\} + 1$, $w = v + 1$, $\hat{\pi}_j = \pi_j$ for $\pi_j < \min\{\pi_r, \pi_s\}$, and $\hat{\pi}_j = \pi_j + 2$ for $\pi_j > \min\{\pi_r, \pi_s\}$. The (g, b)-padding cuts the cycle C into two cycles as follows (see Figure 3).

1. removing edges g and b,
2. adding two new nodes v and w between π_i and π_{i+1},
3. adding two new black edges (π_i, v) and (w, π_{i+1}), and
4. adding two new gray edges (π_r, v) and (w, π_s).

Let ρ be a transposition on three black edges on $\hat{\pi}$. Then ρ can be mimicked by a transposition on π by ignoring the padded elements. In general, let $\rho_1, ..., \rho_t$ be a series of transpositions that sort $\hat{\pi}$ into the identity $\hat{I}$. Then, there are transpositions $\rho'_1, ..., \rho'_t$ that sort π into I.

A black edge (π_i, π_{i+1}) is labeled with $i + 1$. Consider a k-cycle C traversing in order of the black edges $i_1, ..., i_k$. C can be written in k possible ways depending on the choice of the first black edge. We assume that the initial black edge i_1 of C starts at its "leftmost" node in π, i.e., $i_1 = \min_{1 \le t \le k} i_t$. Let $C = (i_1, i_2, ..., i_k)$ be a k-cycle with $k > 3$. We cut C into two cycles C_1 and C_2 by a (g, b)-padding, where g is the gray edge adjacent to the left node π_{i_1-1} of the black edge i_1 $((\pi_{i_1-1}, \pi_{i_1}))$ and b is the black edge i_3. Obviously, C_1 is a 3-cycle and C_2 is a $(k-2)$-cycle (see Figure 3).

For a permutation π, let $\hat{\pi}$ be the permutation obtained from applying the (g, b)-padding on C above. If C is an odd cycle then C_1 and C_2 are odd cycles. If C is an even cycle then C_1 is odd and C_2 is even. From this, $b(\hat{\pi}) = b(\pi) + 1$ and $c(\hat{\pi}) = c(\pi) + 1$ which implies $b(\pi) - c(\pi) = b(\hat{\pi}) - c(\hat{\pi})$. Therefore, given a permutation π with long cycles in $G(\pi)$, we can always transform π into a permutation $\hat{\pi}$ by (g, b)-paddings such that $G(\hat{\pi})$ has only short cycles and $b(\pi) - c(\pi) = b(\hat{\pi}) - c(\hat{\pi})$.

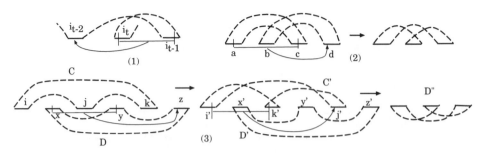

FIGURE 4. A 2-move on an oriented cycle, and removing 2-cycles and interleaved 3-cycles.

A cycle $C = (i_1, i_2, ..., i_k)$ is called *oriented* if there exists $1 < t \le k$ such that $i_{t-1} > i_t$ and either $i_{t-2} \not\subseteq [i_t, i_{t-1}]$ or $i_{t+1} \notin [i_t, i_{t-1}]$ otherwise *non-oriented*. Notice that $i_1 < i_2$ for every cycle. This implies that every 2-cycle is non-oriented.

LEMMA 3.1. *If there is an oriented cycle in $G(\pi)$ then a 2-move is possible in π.*

PROOF. Let $C = (i_1, ..., i_k)$ be an oriented cycle and let $3 \le t \le k$ be the smallest index such that $i_{t-1} > i_t$. Then the transposition $t(i_t, i_{t-1}, i_{t-2})$ eliminates C for $k = 3$ and reduces the length of C by 2 (see (1) of Figure 4). Therefore, $t(i_t, i_{t-1}, i_{t-2})$ is a 2-move. ☐

Ordered sequence of integers $\{v_1 < ... < v_k\}$ and $\{w_1 < ... < w_k\}$ are interleaving if either $v_1 < w_1 < v_2 < w_2 < ... < v_k < w_k$ or $w_1 < v_1 < w_2 < v_2 < ... < w_k < v_k$. For simplicity, we sometimes denote a black edge (π_i, π_{i+1}) by π_{i+1} when no confusion arises. We say a cycle C interleaves with black edges a and b if there are black edges c and d in C so that $a < c < b < d$ or $c < a < d < b$. We say a cycle $C = (i_1, ..., i_k)$ is spanned by a cycle $D = (j_1, ..., j_l)$ if $\min\{j_1, ..., j_l\} < \min\{i_1, ..., i_k\}$ and $\max\{j_1, ..., j_l\} > \max\{i_1, ..., i_k\}$.

LEMMA 3.2. *Given a non-oriented cycle C and arbitrary black edges a and b ($a < b$), there exists a cycle D such that a, b and D interleave.*

PROOF. It is easy to check that there is at least one black edge of a cycle other than C between black edges a and b. Let c be the black edge in a cycle other than C such that π_c is the largest element among the $\pi_i's$ which appear in the black edges of the cycles other than C between a and b. Then $\pi_c + 1$ is not in the interval $[a - 1, b]$. Let $\pi_d = \pi_c + 1$. Then d is a black edge in the cycle which contains c. Obviously black edges a, b and c, d interleave. ☐

THEOREM 3.3. *For any permutation π such that $G(\pi)$ has only short cycles, either a 2-move or a 0-move followed by two consecutive 2-moves are possible in π.*

PROOF. If $G(\pi)$ has an oriented cycle then from Lemma 3.1 a 2-move is possible. If $G(\pi)$ has 2-cycles only then let $C = (a, b)$ be a 2-cycle. From Lemma 3.2, there exists a 2-cycle $D = (c, d)$ such that (a, b) and (c, d) interleave. Assume that $a < c < b < d$. Then the transposition $t(a, c, d)$ merges the two cycles C and D into one oriented 3-cycle (see (2) of Figure 4). Since $t(a, c, d)$ reduces the number of breakpoints by one and increases the number of odd cycles by one, $t(a, c, d)$ is a 2-move. In fact, π can be sorted into I by a sequence of consecutive 2-moves in this case.

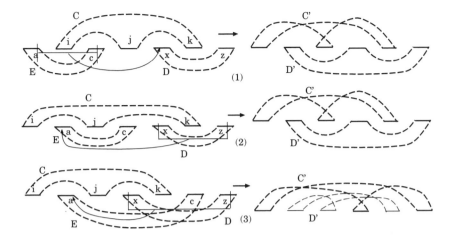

FIGURE 5. Case 1 of Theorem 3.3.

Now, assume that $G(\pi)$ does not have any oriented cycle and has at least one non-oriented 3-cycle. Let $C = (i, j, k)$ be a non-oriented 3-cycle. If there is a cycle $D = (x, y, z)$ such that (i, j, k) and (x, y, z) interleave, say $i < x < j < y < k < z$, then transposition $\rho = t(x, y, z)$ transforms the cycles C and D into cycles $C' = (i', j', k')$ and $D' = (x', y', z')$, where (i', j', k') and (x', y', z') interleave, and C' is oriented (see (3) of Figure 4). ρ is a 0-move. Since C' is oriented, $t(i', k', j')$ is a 2-move which removes C' and transforms D' into an oriented cycle D''. Then we can find another 2-move to remove D''.

So, we assume that no pair of 3-cycles interleave. Let $C = (i, j, k)$ be a 3-cycle which is not spanned by any other 3-cycle. For the pair (i, k), from Lemma 3.2, there is a cycle D which contains black edges r and s such that (i, k) and (r, s) interleave. We assume that $i < r < k < s$ (the case of $r < i < s < k$ can be proved symmetrically). Let x be the minimum edge and z the maximum edge of D. Since C is not spanned by D, $i < x < k < z$. The proof is now partitioned into two subcases.

Case 1: $j < x < k < z$.

Find a cycle E which interleaves with (i, j). Let a be the minimum edge and c the maximum edge of E. Then either $a < i < c < j$ or $i < a < j < c$ (otherwise either $E = (a, b, c)$, (a, b, c) and (i, j, k) interleave; or C is spanned by E). For the case of $a < i < c < j$, transposition $t(a, c, x)$ transforms C into an oriented cycle C' and merges cycles D and E into one cycle D' (see (1) of Figure 5). Notice that $t(a, c, x)$ either reduces the number of odd cycles by one or increases the number by one. Also $t(a, c, x)$ removes one break point. Therefore, $t(a, c, x)$ is either a 0-move or a 2-move. After transposition $t(a, c, x)$, we can find a 2-move which removes C' and transforms D' into an oriented cycle D''. From Lemma 1, another 2-move on D'' is possible.

The case of $i < a < j < c$ is further partitioned into following subcases based on the value of c: $c < x$, $x < c < k$, $k < c < z$, and $z < c$. For all the four subcases, transposition $t(x, z, a)$ transforms C into an oriented cycle C' and merges cycles D and E into one cycle D'. For the cases of $c < x$ and $z < c$, two 2-moves can be found as shown in the case of $a < i < c < j$ (see (2) of Figure 5). For the cases of $x < c < k$ and $k < c < z$, we get two oriented cycles as shown in (3) of Figure 5, and two 2-moves are possible.

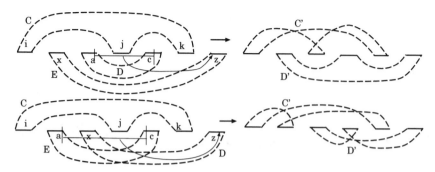

FIGURE 6. Case 2 of Theorem 3.3.

Algorithm Transposition_Sort(π);
begin
    Construct $G(\pi)$ and let $C_1, ..., C_r$ be the disjoint cycles of $G(\pi)$;
    Cut long cycles in $G(\pi)$ into short ones.
    while ($\exists C_i$) **do**{
        while ($\exists C_i$, C_i is oriented) **do**
            {remove C_i by one transposition;}
        if $\exists$ $C = (i, j, k)$ and $C' = (x, y, z)$ s.t. (i, j, k) and (x, y, z) interleave **then**
            remove C and C' by three transpositions;
        else find a 2-move, or a 0-move and two 2-moves as shown in Theorem 3.3;
    }
end.

FIGURE 7. Sorting permutations by transpositions.

Case 2: $i < x < j$ and $k < z$.
Find a cycle E which interleaves with (j, k). Let a be the minimum edge and c the maximum edge of E. Then either $a < j < c < k$ or $j < a < k < c$. The case of $j < a < k < c$ has been proved in Case 1. In the case of $a < j < c < k$, there are three subcases: $x < a$, $i < a < x$, and $a < i$. For all the three subcases, transposition $t(a, c, z)$ transforms C into an oriented cycle C' and merges cycles D and E into one cycle D'. Following a similar argument as that in Case 1, two 2-moves are possible after $t(a, c, z)$ (see Figure 6). $\square$

Now, we are ready to give our algorithm (see Figure 7).

THEOREM 3.4. *Algorithm Transposition_Sort is a 1.5-approximation algorithm for sorting permutations by transpositions. The time complexity of the algorithm is $O(c(\hat{\pi})n)$ for a permutation π of n elements.*

PROOF. Theorem 3.3 guarantees that we can reduce the measure $b(\hat{\pi}) - c(\hat{\pi})$ by at least four in three transpositions. From this, we can sort $\hat{\pi}$ by at most $(3/4)(b(\hat{\pi}) - c(\hat{\pi}))$ transpositions. Since $b(\hat{\pi}) - c(\hat{\pi}) = b(\pi) - c(\pi)$, we can sort π by at most $(3/4)(b(\pi) - c(\pi))$ transpositions. From the lower bound $(b(\pi) - c(\pi))/2$, algorithm Transposition_Sort is a 1.5-approximation one.

The padding procedure can be done in linear time by assigning the new inserted nodes v and w real numbers and relabeling the values of all the node in $G(\pi)$ after all the long cycles have been cut. It takes $O(c(\hat{\pi})n)$ time to remove the 3-cycles in $G(\hat{\pi})$. Removing

2-cycles takes linear time. Thus, the time complexity of algorithm Transposition_Sort is $O(c(\hat{\pi})n)$. □

4. Sorting permutations by reversals, transpositions, and reversals+transpositions

Gu, Peng, and Sudborough studied the problem of sorting signed permutations by reversals, transpositions, reversal+transpositions and gave a 2-approximation algorithm for the problem [**GPS96**]. Now, we propose a heuristic algorithm for the same problem. The algorithm follows a greedy strategy which always executes an i-move with the maximum i. From the lower bound stated in the previous section, an operation of reversal, transposition, or reversal+transposition is an i-move with $i \leq 2$. The algorithm checks the cycles, one by one, to find a 2-move. For a signed permutation π, Figure 8 gives the cases of the cycles in the breakpoint graph $G(\pi)$ such that a 2-move is possible in π. The algorithm is described

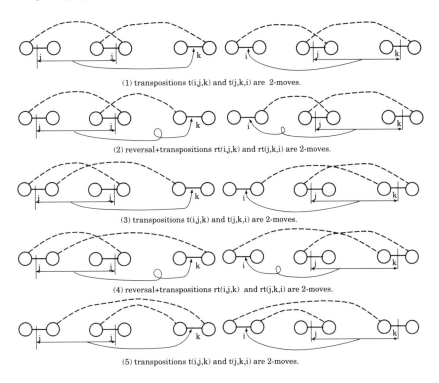

(1) transpositions t(i,j,k) and t(j,k,i) are 2-moves.

(2) reversal+transpositions rt(i,j,k) and rt(j,k,i) are 2-moves.

(3) transpositions t(i,j,k) and t(j,k,i) are 2-moves.

(4) reversal+transpositions rt(i,j,k) and rt(j,k,i) are 2-moves.

(5) transpositions t(i,j,k) and t(j,k,i) are 2-moves.

FIGURE 8. The 2-moves.

in Figure 9. The time complexity of the algorithm is $O(n^3)$.

We tested the heuristic algorithm on randomly generated permutations. We used three classes of permutations: (1) the permutations $\pi = (\pi_1 \pi_2 ... \pi_n)$ in which each π_i is generated uniformly from $[1, n]$ exclusively, (2) permutations obtained by applying $\sqrt{n}$ randomly generated operations (reversal, transposition, and reversal+transposition are generated with the equal probability of 1/3) to I, and (3) permutations obtained by applying $\log_2 n$ randomly generated operations to I. For each class of permutations, we used five samples to test the algorithm. Table 1 gives average values of the lower bound (given by $(b(\pi) - c(\pi))/2$) and the number of operations (upper bound) of sorting the permutations by the algorithm for class (1). Tables 2 and 3 give the average values for classes (2) and

Algorithm SORT(π);
begin
    Let $C_1, ..., C_r$ be the cycles of $G(\pi)$;
    while ($\exists C_i$) **do**{
        if ($\exists$ a 2-move in π) **then**
            perform the 2-move
        else perform a 1-move or 0-move;
    }
end.

FIGURE 9. The heuristic algorithm.

n	200	300	400	500	600	700	800	900	1000
Lower bound	99.4	149.4	199	249.6	299	350	398.8	449.4	499.2
Upper bound	100.6	150.6	200.8	251	300	351	399.6	450.4	500

TABLE 1

n	200	300	400	500	600	700	800	900	1000
Lower bound	13.6	16.8	20	22	23.8	25.6	28	30	30.8
Upper bound	14.8	18.4	21.4	24.6	26.2	26.8	29.6	33.4	34.4

TABLE 2

n	200	300	400	500	600	700	800	900	1000
Lower bound	8.8	15	13	11.2	13.4	18.4	11.8	13.4	8
Upper bound	8.8	16.4	13.8	11.8	13.6	19	12.2	14	8.6

TABLE 3

(3), respectively. The experiment results show that the algorithm sorts randomly generated permutations in an almost optimal number of operations.

5. Conclusional Remarks

Computational approaches provide efficient tools for large-scale comparative genetic mapping which offers exciting prospects for understanding the evolution of genomes. This paper proposed algorithms for computing the distance between genomes in the sense of global (reversals/transpositions) rearrangements. We transformed the genome rearrangements problem into the problem of sorting a permutation. We proposed a 1.5-approximation algorithm for sorting unsigned permutations by transpositions only. Our algorithm is much simpler than the previous one. We also gave a heuristic algorithm for sorting signed permutations by reversals/transpositions. The heuristic algorithm sorts randomly generated permutations in almost optimal number of operations. Future work includes reducing the error bounds of the approximation algorithm further and applying the algorithms to practical biology data.

Acknowledgements

Mr. Kazuyuki Iwata tested the heuristic algorithm. This work was partially supported by the grant-in-aid for scientific research on the priority area "Genome Science" from the Ministry of Education, Science, Sports and Culture of Japan.

References

[BP95]   V. Banfa and P. Pevzner, Sorting permutations by transpositions, *Proc. of 6th ACM-SIAM Annual Symposium on Discrete Algorithms*, 1995, pp. 614–623.

[BP96]   V. Bafna and P. Pevzner, Genome rearrangements and sorting by reversals, *SIAM J. on Computing* **25** (1996), no. 2, 272–289.

[Cap97] A. Caprara, Sorting by reversals is difficult, *Proc. of the 1st Annual International Conference on Computational Molecular Biology*, 1997, pp. 75–83.

[GPS96] Q.P. Gu, S. Peng, and H. Sudborough, A 2-approximation algorithm for genome rearrangements, *Theoretical Computer Science*, to appear, 1998.

[HP95]   S. Hannenhalli and P. Pevzner, Transforming cabbage into turnip (polynomial algorithm for sorting signed permutation by reversals), *Proc. of 27th ACM Symposium on Theory of Computing (STOC'95)*, 1995, pp. 178–189.

[KS93]   J. Kececioglu and D. Sankoff, Exact and approximation algorithms for the inversion distance between two permutations, *Proc. of the 4th Annual Symposium on Combinatorial Pattern Matching, Lecture Notes in Computer Science* **684**, 1993, pp. 87–105 (Extended version has appeared in *Algorithmica* 13:180–210, 1995).

[KST97] H. Kaplan, R. Shamir, and R.E. Tarjan, Faster and simpler algorithm for sorting signed permutations by reversals, *Proc. of the 8th ACM-SIAM Annual Symposium on Discrete Algorithms*, 1997, pp. 344–351.

[San92] D. Sankoff, Edit distance for genome comparison based on non-local operations, *Lecture Notes in Computer Science* **644**, 1992, pp. 121–135.

[SCA90] D. Sankoff, R. Cedergren, and Y. Abel, Genomic divergence through gene rearrangement, *Methods in Enzymology* **183**, 1990, pp. 428–438.

UNIVERSITY OF AIZU, AIZU-WAKAMATSU, FUKUSHIMA, JAPAN 965-8580
E-mail address: qian@u-aizu.ac.jp

UNIVERSITY OF AIZU, AIZU-WAKAMATSU, FUKUSHIMA, JAPAN 965-8580
E-mail address: s-peng@u-aizu.ac.jp

UNIVERSITY OF AIZU, AIZU-WAKAMATSU, FUKUSHIMA, JAPAN 965-8580
E-mail address: qmchen@u-aizu.ac.jp